Environmental Policies and NGO Influence

Environmental Policies and NGO Influence examines why non-governmental organisations (NGOs) are at times able to exert influence on policies to conserve and use natural resources sustainably in sub-Saharan Africa. The book is unique in bringing together NGO campaigners and researchers in three African countries with academics specialising in development studies, systems and environmental policy at the Open University.

The two opening chapters provide essential background material on the conceptual framework of the study and on land resource issues in sub-Saharan Africa. The next seven chapters are devoted to case studies of NGO activity. The book concludes with a summary of the lessons the studies yield for NGO campaigners and policy specialists.

The first case study examines NGO influence on the negotiations for the UN Convention to Combat Desertification. It is followed by two case studies each from three African countries: Zimbabwe, Botswana and Nigeria. Three chapters examine campaigns on land use policy, namely opposition to the fencing of communal rangelands in Botswana, the promotion of small grains to challenge the market dominance of maize in Zimbabwe and the conflicts between local communities and the Shell oil company in the Niger delta. Three chapters on water use issues cover the campaign for water from the Zambezi for Bulawayo, the campaign against a planned development of the Okavango Delta in Botswana and the campaign against the Kafin Zaki Dam in Nigeria.

The book illustrates the complex nature of governance and policy making for resource issues with international, national and local dimensions. It is essential reading for environment and development policy practitioners and for students of development studies, African politics and international relations.

Alan Thomas is Senior Lecturer in the Development Policy and Practice discipline at the Open University. **Susan Carr** is Senior Lecturer in Systems specialising in agricultural and environmental management at the Open University. **David Humphreys** is Lecturer in Environmental Policy specialising in forests politics at the Open University.

Routledge Research Global Environmental Change Series

1 **The Condition of Sustainability**
Ian Drummond and Terry Marsden

2 **Global Trade and Global Social Issues**
Edited by Annie Taylor and Caroline Thomas

3 **Environmental Policies and NGO Influence**
Land degradation and sustainable resource management in Sub-Saharan Africa
Edited by Alan Thomas, Susan Carr and David Humphreys

4 **The Sociology of Energy, Buildings and the Environment**
Constructing knowledge, designing practice
Simon Guy and Elizabeth Shove

5 **Living with Environmental Change**
Social vulnerability, adaptation and resiliance in Vietnam
Edited by W. Neil Adger, P. Mick Kelly and Nguyen Huu Ninh

6 **Land and Limits**
Interpreting sustainability in the planning process
Susan Owens and Richard Cowell

Already published in the Routledge Global Environmental Change Series

Timescapes of Modernity
Barbara Adam

Reframing Deforestation
James Fairhead and Melissa Leach

British Environmental Policy and Europe
Edited by Philip Lowe

The Politics of Sustainable Development
Edited by Susan Baker, Maria Kousis, Dick Richardson and Stephen Young

Argument in the Greenhouse
Nick Mabey, Stephen Hall, Clare Smith and Sujata Gupta

Environmentalism and the Mass Media
Graham Chapman, Keval Kumar, Caroline Fraser and Ivor Gaber

Environmental Change in Southeast Asia
Edited by Michael Parnwell and Raymond Bryant

The Politics of Climate Change
Edited by Timothy O'Riordan and Jill Jager

Population and Food
Tim Dyson

The Environment and International Relations
Edited by John Vogler and Mark Imber

Global Warming and Energy Demand
Edited by Terry Barker, Paul Ekins and Nick Johnstone

Social Theory and the Global Environment
Michael Redclift and Ted Benton

Working the Sahel
Environment and society in Northern Nigeria
Michael Mortimore and Bill Adams

Environmental Policies and NGO Influence

Land degradation and sustainable resource management in sub-Saharan Africa

Edited by Alan Thomas, Susan Carr and David Humphreys

Global Environmental Change Programme

ROUTLEDGE
Taylor & Francis Group

London and New York

First published 2001
by Routledge
11 New Fetter Lane, London EC4P 4EE

Simultaneously published in the USA and Canada
by Routledge
29 West 35th Street, New York, NY 10001

Routledge is an imprint of the Taylor and Francis Group

Typeset in Sabon by Helen Skelton, London
Printed and bound in Great Britain by Biddles Ltd,
Guildford and King's Lynn

British Library Cataloguing in Publication Data
A catalogue record for this book is available from the British Library

Library of Congress Cataloging in Publication Data
Environmental policies and NGO influence: land degradation and sustainable resource management in sub-Saharan Africa / edited by Alan Thomas, Susan Carr, and David Humphreys.
p. cm – "Global Environmental Change Programme."
Includes bibliographical references and index.
1. Sustainable development–Africa, sub-Saharan. 2. Environmental policy–Africa, sub-Saharan. 3. Non-governmental organizations–Africa, sub-Saharan. 4. Land use–environmental aspects–Africa, sub-Saharan. 5. Water use–environmental aspects –Africa, sub-Saharan. 6. Conservation of natural resources–Africa, sub-Saharan.
I. Thomas, Alan. II. Carr, Susan.
III. Humphreys, David.
HC800.Z9 E5398 2000
333.7'0967–dc21 00-025632

ISBN 0-415-18963-2

For Timothy

Contents

Figures and tables

Figures

Tables

Contributors

Susan Carr is Senior Lecturer in Systems, Faculty of Technology, Open University

Oronto Douglas is Chief Field Officer, Environmental Rights Action, Nigeria

Derek Gunby is Director, PlanAfric Consultancy, Bulawayo, Zimbabwe

David Humphreys is Lecturer in Environmental Policy, Faculty of Social Sciences, Open University

Roger Mpande is Co-Director of the Community Technology Development Association (COMMUTECH), Harare and a member of the Zimbabwe Environmental Research Organisation (ZERO)

Obasi Ogbonnaya is Editor, Change Radio Nigeria; formerly Campaigns and Treaties Manager for the Nigerian Conservation Foundation, and journalist on the Nigerian *Guardian*

Enoch Okpara is Director of the Nigerian Environmental Study/Action Team (NEST) and Professor of Environmental Sciences, Imo State University

Uche Onyeagucha is Chief Campaigner, Environmental Rights Action, Nigeria

Onalenna Selolwane is Senior Lecturer and Head of Sociology Department, University of Botswana

Alan Thomas is Senior Lecturer in the Development Policy and Practice discipline, Faculty of Technology, Open University

Acknowledgements

The editors acknowledge the support of the Economic and Social Research Council (ESRC) which funded the Global Environmental Change Open University (GECOU) research project. We are also grateful to the participants in the international seminar on *Environmental Policies and NGO Influence: Land Degradation and Resource Management* held in Harare in June 1996 (Annex); their constructive input helped to shape the framework and contents of this book. The seminar was generously sponsored by the ESRC, the British Council and the Swedish International Development Authority. Many thanks to Roger Mpande and the staff at COMMUTECH, Harare for their help in organising the seminar; to Andrew Blowers, Bernard Eccleston, Dick Morris and David Potter who commented on earlier drafts of the papers in this volume; to John Hunt for the illustrations; to Cathy McNulty for helping to finalise the manuscript; and to the many people and organisations who generously assisted us in our research.

Acronyms, abbreviations and organisations

Agritex	Agricultural Extension Service (Zimbabwe)
BDP	Botswana Democratic Party
BNF	Botswana National Front
BOT	build, operate and transfer
CAMPFIRE	Communal Areas Management Programme for Indigenous Resources (Zimbabwe)
CCD	Convention to Combat Desertification
CCF	Centre for our Common Future
CLO	Civil Liberties Organization (Nigeria)
COC	Council of Ogoni Churches (Nigeria)
COP	Council of Ogoni Professionals (Nigeria)
COTRA	Council of Ogoni Traditional Rulers (Nigeria)
CSD	United Nations Commission on Sustainable Development
CST	Committee on Science and Technology (of the INCD/CCD)
CZI	Congress of Zimbabwean Industry
DEAP	District Environmental Action Plan (Zimbabwe)
DNR	Department of Natural Resources (Zimbabwe)
DWD	Department of Water Development (Zimbabwe)
EC/EU	European Community/ European Union
ECOSOC	United Nations Economic and Social Council
EIA	Environmental Intelligence Agency
EIA	environmental impact assessment
ELCI	Environment Liaison Centre International
ENDA	Environment and Development Activities
ENDA-ZW	Environment and Development Activities – Zimbabwe
ERA	Environmental Rights Action (Nigeria)
ESAP	Economic Structural Adjustment Programme (Zimbabwe)
FAB	Forestry Association of Botswana
FAO	United Nations Food and Agriculture Organization
FEPA	Federal Environmental Protection Agency (Nigeria)
FEWA	Federation of Ogoni Women's Associations (Nigeria)

FRN	Federal Republic of Nigeria
FONSAG	Forum for Sustainable Agriculture (Botswana)
FORMECU	Forestry Management, Evaluation and Co-ordinating Unit (Nigeria)
G77	Group of 77 (developing countries)
GNP	gross national product
GMB	Grain Marketing Board (Zimbabwe)
GRO	grassroots organization
GRSO	grassroots support organization
HNWCP	Hadejia-Nguru Wetlands Conservation Project (Nigeria)
ICRISAT	International Crop Research Institute for the Semi-Arid Tropics
IDRC	International Development and Research Centre
IFAD	International Fund for Agricultural Development
IGO	inter-governmental organization
IIED	International Institute for Environment and Development
IISD	International Institute for Sustainable Development
IMF	International Monetary Fund
INCD	International Negotiating Committee for a Convention to Combat Desertification
IUCN	World Conservation Union (formerly International Union for the Conservation of Nature and Natural Resources)
KAGOTE	umbrella organization for Ogoni traditional rulers (Nigeria)
KCS	Kalahari Conservation Society (Botswana)
KENGO	Kenya Energy and Environment Organizations
KRIP	Kano River Irrigation Project (Nigeria)
MDA	Manicaland Development Association (Zimbabwe)
MOSOP	Movement for the Survival of the Ogoni People (Nigeria)
MP	Member of Parliament
MZWP	Matabeleland Zambezi Water Project (Zimbabwe)
MZWT	Matabeleland Zambezi Water Trust (Zimbabwe)
NACSD	National Advisory Committee on Sustainable Development (Zimbabwe)
NCCD	national NGO co-ordinating committee on desertification
NCF	Nigerian Conservation Foundation
NCS	National Conservation Strategy
NEST	Nigerian Environmental Study/Action Team
NIPSS	National Institute for Policy and Strategic Studies (Nigeria)
NGO	non-governmental organization
NNPC	Nigerian National Petroleum Corporation
NORAD	Norwegian Agency for International Development

NOVIB	Netherlands Organization for International Development Co-operation
NPA	New Policy Agenda
NUOS	National Union of Ogoni Students (Nigeria)
NYCOP	National Youth Council of Ogoni People (Nigeria)
OCU	Ogoni Central Union (Nigeria)
OECD	Organization for Economic Co-operation and Development
OMPADEC	Oil Mineral Producing Area Development Commission (Nigeria)
ORAP	Organization of Rural Associations for Progress (Zimbabwe)
OSU	Ogoni Students Union (Nigeria)
OTU	Ogoni Teachers Union (Nigeria)
OWLS	Okavango Wild Life Society (South Africa)
PIRC	Pensions and Investments Research Consultants Ltd
ROAPE	Review of African Political Economy
RIOD	Réseau International d'ONG sur la Désertification (International NGO Network on Desertification)
RSPB	Royal Society for the Protection of Birds
SADC	Southern African Development Community
SANDON	Southern Africa Network of Development Organizations
SAP	structural adjustment programme
SIDA	Swedish International Development Authority
SMEC	Snowy Mountains Engineering Corporation
SMIP	Sorghum and Millet Improvement Programme (of ICRISAT)
SNV	Netherlands Development Organization
SPDC	Shell Petroleum Development Company of Nigeria
SOIWDP	Southern Okavango Integrated Water Development Project (Botswana)
TGLP	Tribal Grazing Lands Policy (Botswana)
TNC	transnational corporation
TOCT	Tshomorelo Okavango Conservation Trust (Botswana)
UIA	Union of International Associations
UK	United Kingdom
UN	United Nations
UNCED	United Nations Conference on Environment and Development
UNDP	United Nations Development Programme
UNEP	United Nations Environment Programme
UNPO	Unrepresented Nations and Peoples Organization
UNSO	United Nations Sudano-Sahelian Office (now the UN Office to Combat Desertification and Drought)
US/USA	United States/United States of America

USAID	United States Agency for International Development
WRI	World Resources Institute
WTO	World Trade Organization
WWF	World Wide Fund for Nature
ZANU	Zimbabwe African National Union
ZANU-PF	Zimbabwe African National Union – Patriotic Front
ZAPU	Zimbabwe African People's Union
ZERO	originally Zimbabwe Environmental Research Organization; now 'regional network of environmental experts'
ZFU	Zimbabwe Farmers Union
ZIMBAC	Zimbabwe Biotechnology Advisory Committee
ZSAN	Zimbabwe Seeds Action Network

1 NGOs and their influence on environmental policies in Africa

A framework

Alan Thomas

Why are non-governmental organisations (NGOs) influential in affecting the development of certain policies related to global environmental problems?

The contributions in this volume address this question by presenting case study evidence from three sub-Saharan African countries, namely Nigeria, Zimbabwe and Botswana. NGO influence is examined at the international level (at negotiations for the Convention to Combat Desertification) and also at the national and local levels with examples of environment, land use and water development policies.

An NGO is defined here as not only non-governmental, but also non-profit, and as an organisation it has 'at least several full-time staff, some sort of hierarchy, a budget and an office' (Potter and Taylor 1996: 1). Environmental policy is defined as 'a deliberate course of action by a target organization, or set of organizations, designed to accomplish some end related to an environmental problem' (Potter and Taylor 1996: 2–3). An environmental problem is 'any change in state in the physical environment which is brought about by human interference with the physical environment, and has effects which society deems unacceptable in the light of its shared norms' (Sloep and van Dam-Mieras 1995: 42; Potter and Taylor 1996: 2). An environmental problem may be said to be 'global' when its causes are global, its effects are global, or the actors involved transcend state boundaries.

Why are non-governmental organisations (NGOs) able to influence some environmental policies but not others?

There are two purposes behind analysing examples of NGO influence. One is to help NGOs think more clearly about how to improve their campaigning strategies. Since many of the factors likely to determine NGO influence are outside their control – for example, where NGOs oppose entrenched political or economic interests – the scope for this may appear to be limited. However, the case studies do illustrate how NGOs have used a range of strategies to greater or lesser effect in a variety of circumstances. A broader purpose is to find a way of understanding NGOs' changing role in the search for sustainable development.

NGOs and their changing role

NGOs are of increasing importance at all levels: local, national and global. At the international level, the number of NGOs listed in the Yearbook of International Organizations (UIA 1996) has risen from just over 1,000 in the late 1950s to over 20,000 in the 1990s, though Willetts (1996) suggests that only about a quarter of these are truly international. At the other end of the spectrum, Fisher (1998) estimates at over 200,000 grassroots organisations (GROs) in Asia, Africa and Latin America, plus at least 50,000 grassroots support organisations (GRSOs) (Fisher 1993; UNDP 1993).

The number of NGOs with consultative status with the United Nations Economic and Social Council (ECOSOC) increased steadily from about 200 in 1950 to almost 1,000 in the early 1990s and 1,500 by 1995 (Willetts 1996: 38). Willetts also claims that in world politics NGOs 'dominate agenda-setting' (1996: 46). It is generally accepted that environmental NGOs are largely responsible for environmental interests coming to the fore internationally. For example, while there was no mention of the environment in the original UN charter, the United Nations Environment Programme (UNEP) and the Commission on Sustainable Development (CSD) are now major UN organs (Morphet 1996). The development arena has also changed considerably due to increasing NGO participation. By 1989 over 4,000 development NGOs based in OECD countries spent over US$6 billion annually on overseas aid, working with between 10,000 and 20,000 Southern NGOs who assisted approximately 100 million people (Clark 1991: 47, 51). By the mid-1990s this annual figure was over US$7 billion, including over US$1 billion of official aid channelled through NGOs, representing for some countries more than 10 per cent of all such aid (OECD 1998). According to Fisher (1998: 7) the number of people assisted by NGOs was more like 300 million.

The impact of NGOs therefore extends beyond their increasing numbers. Governments and international agencies now afford them a broader role. Edwards and Hulme explain how a 'New Policy Agenda' (NPA) has emerged combining market economics and liberal democratic politics, in which NGOs are simultaneously 'viewed as market-based actors' and 'placed … in a central position as components of "civil society"' (1995: 849). Thus the increase in provision of services by NGOs, including those with environmental relevance such as agricultural extension and marketing, is seen as part of the privatisation of state services, despite NGOs' non-profit basis. This is often described as 'gap-filling' (e.g. Vivian 1994), and occurs where states especially in the South have insufficient capacity to maintain universal services or are obliged to reduce them as part of a structural adjustment or donor-promoted 'reform' package.

The NPA also gives NGOs a prime role in democratisation (Clark 1991) and the development of 'civil society' (Van Rooy 1998). ROAPE argues that democracy requires 'a widespread and complex process involving the strengthening of civil society' and NGOs are 'one of the institutional forms that can deepen [civil society]' (1992: 3–4). However, Fowler (1993) warns against assuming that NGOs act automatically as agents of democratisation. In global civil society too NGOs are often credited with a leading role, either as a counter to the multilateral financial institutions and UN agencies and conventions, or potentially alongside them (Princen and Finger 1994; Wapner 1995; Fox and Brown 1998).

Characteristics of NGOs

As *organisations* which are non-governmental and non-profit, NGOs are only a part of civil society. Social networks and social movements are not NGOs, though the latter are of increasing importance throughout the South (Wignaraja 1993) and NGOs may provide an organisational focus for such movements. Conversely, some NGOs at least are 'embedded in the dynamic and the interests of popular groups and movements' (Landim 1993: 219). Most writers restrict usage of the term 'NGO' further, excluding political parties and religious congregations and, in some cases, peasants' organisations and trade unions. Some, such as Uphoff (1993), differentiate grassroots organisations (GROs) from NGOs, though in this book we consider organisations at all levels – international, national and local. We also include membership organisations as well as those oriented towards social or public benefit.

NGOs tend to be conceived differently in different sectors. In particular, their role and even what they are is seen differently for environmental and development NGOs. The major environmental NGOs are primarily membership organisations based in the North which promote and lobby for environmental interests nationally and internationally. Development NGOs began as charitable relief or missionary welfare organisations, often running localised projects to alleviate rural poverty (Riddell and Robinson 1995).

Korten (1987) has suggested that development NGOs have typically moved on to the promotion of small-scale self-reliance and then to a 'third generation strategy' of 'sustainable systems development' involving 'facilitating development by other organizations, both public and private' (1987: 149), though emergency relief is still an important part of their work. Edwards and Hulme point out that, despite many small-scale successes, the impact of development NGOs is 'highly localised, and often transitory' (1992: 13). They suggest that the pursuit of 'making a difference' has led development NGOs to try a variety of methods of 'scaling up' (Annis 1988) including not only 'additive' strategies where the size of programmes is increased, but also 'multiplicative' strategies involving 'deliberate influence, networking, policy and legal reform' and 'diffusive' strategies spreading ideas informally and spontaneously (Edwards and Hulme 1992: 15).

Thus the role of development NGOs has broadened to include advocacy alongside the provision of services. At the same time environmental NGOs may provide development services that complement their promotional and lobbying work. For example, in Zimbabwe the World Wide Fund for Nature (WWF) actively supports the 'CAMPFIRE' programme for the sustainable utilisation of natural resources (particularly wildlife) at community level. In Nigeria the Royal Society for the Protection of Birds (RSPB) has supported a project to promote community development alongside its bird protection activities (Chapter 8).

There has been a huge growth in Southern NGOs, especially grassroots organisations (GROs) based on solidarity amongst a membership in particular poor communities (Ghai and Vivian 1992; Fisher 1993, 1998). The interests of such communities tend to centre on questions of livelihood and do not distinguish between environment and development. Many of the Southern-based grassroots support organisations (GRSOs) which have grown in support of these GROs have become NGOs with a significant national and in some cases global lobbying presence.

Types of NGO in our case study contexts

The potential for variety amongst NGOs is enormous. For example, one type of NGO has other NGOs as members. Examples include the various representative bodies or fora which exist in most countries as well as international federations such as Friends of the Earth International. We might regard international institutes such as the World Resources Institute in Washington, the International Institute for Environment and Development in London, or the World Watch Institute, as another type, characterised by small, professional expert staff dedicated to research and consultancy on environmental and development issues. However, the main types of NGO active in Africa are outlined below.

In Africa, as elsewhere, the NGOs which now seek to influence environmental policy have started from very different roots. This is in line with Handy's observation of British voluntary organisations that, although they start as one of three categories, namely mutual support, service delivery or campaigning, most end up as an 'amalgam of all three' (1988: 18). However, the different histories, cultural origins and underlying values of NGOs lead to very different involvements in environmental policy. NGOs differ also in the interest groups they represent, either directly for member organisations or indirectly when a charitable or welfare organisation represents the interests of client groups. Hence there is not always a unified NGO view; different NGOs may put forward different views based on conflicting interests.

Several writers have suggested principles for categorising NGOs (e.g. Riddell 1990; Uphoff 1993; Fisher 1993). Following Handy (1988) I have distinguished elsewhere (Thomas 1992) between *mutual benefit*

Table 1.1 Types of NGO with examples from the book's case studies

	Mutual benefit	Public benefit	
	Membership NGOs	Campaigning or research NGOs	Charitable or service-providing NGOs
International	*[Federations of local and national membership organisations, e.g. International Co-operative Alliance]*	Environmental NGOs such as Greenpeace (Ch. 7), ENDA (Ch. 4)	Development NGOs such as SNV, Lutheran World Federation (Ch. 4)
National	*[National membership organisations or federations of GROs. ORAP (see below) could be considered here.]*	Conservationist NGOs and pressure groups such as KCS (Ch. 6 & 7), NCF (Ch. 8), MZWP (Ch. 5)	Service-providing NGOs such as Silveira House (Ch. 4), Thusano Lefatsheng (Ch. 6)
Local	Member-based 'grassroots organisations' (GROs) and local environmental campaigns such as ORAP (Ch. 4), TOCT (Ch. 7), MOSOP (Ch. 9)	*[Local campaigns for public benefit. TOCT (see left) could be considered here but is also a member organisation]*	*[Local charities or service-providing NGOs – not important in our case studies]*

Key

ENDA	Environment and Development Activities
SNV	Netherlands Development Organization
KCS	Kalahari Conservation Society
NCF	Nigerian Conservation Foundation
MZWP	Matabeleland Zambezi Water Project
ORAP	Organization of Rural Associations for Progress
TOCT	Tshomorelo Okavango Conservation Trust
MOSOP	Movement for the Survival of the Ogoni People

organisations, based on membership, and *public benefit* organisations. The latter include charitable and other NGOs which supply services to others together with campaigning or research organisations which have no defined client group. It is also important to recognise differences in scope and scale; how NGOs work at local, national and international levels will vary tremendously.

Table 1.1 summarises these distinctions, with examples of particular importance to environmental policy in sub-Saharan Africa. However, their historical background and development as movements has led to significant differences between countries in the relative importance of the types and the interests represented (Thomas 1995). In particular, Nigeria, Zimbabwe and Botswana differ considerably as to which types of NGO are dominant.

Starting at the international level, we can distinguish between campaigning environmental NGOs and charitable or service-providing development NGOs. Both of these are usually based in the North. Mutual benefit organisations based on membership also exist at the international level. Some of these have African members, although they are not of great importance to environmental policies in Africa.

Thus international environmental NGOs are the first type, epitomised by Greenpeace. Since Africa's environmental problems are not of the highest priority in global terms, there may be no day-to-day local involvement. Many such NGOs become involved in environmental policy issues in Africa only from a distance. Thus, although Chapter 7 covers a case of a brief involvement by Greenpeace in Botswana, that organisation has no staff permanently allocated to African issues. By contrast, the World Conservation Union (IUCN) has branches throughout Africa. IUCN has governments as well as NGOs and other institutions as affiliates, and has been called an 'iquango' (international quasi autonomous non-governmental organisation) (Willetts 1990: 275–6). However, the IUCN secretariat operates in much the same way as an international environmental NGO.

There are also a few Southern-based international NGOs. ENDA-Zimbabwe is perhaps the most notable, Zimbabwe being one of eleven countries with a local office of ENDA-Tiers-Monde, established following the 1972 UN Stockholm Conference on the Human Environment. ENDA stands for 'Environment and Development Activities'. ENDA-Zimbabwe has become a substantial NGO in its own right, engaging in advocacy and research as well as local development projects (Chapter 4). Like the other NGOs of this type, ENDA-Zimbabwe has no local membership, though it is committed to a community-based participative methodology.

The second type are Northern-based development agencies such as OXFAM, the Lutheran World Federation, and so on. These usually have branches in the African countries in which they work. Most of these NGOs are engaged in rural development projects. Over the past twenty years they have either added environmental aspects to their projects or simply found such aspects to be an integral part of their work.

In Africa projects by this type of NGO have mushroomed, especially since the 1980s. Seen as independent, private providers, such NGOs fit in with the neo-liberal view of the world in which state provision should be cut back as much as possible. However, their importance varies considerably from country to country depending on the political context and the capacity of the state. In some cases, such as Mozambique, NGOs have been accused of effectively setting up parallel administrations, undermining state control and acting as agents of Northern recolonisation (Hanlon 1991).

There are some similarities with Zimbabwe, where immediately after independence NGOs provided a way for Northern humanitarian interests to work in that country while avoiding direct partnership with the radical nationalist government. As far as the Zimbabwean government was concerned, NGOs brought in resources, provided services where the state could not fulfil all the expectations built up by the successful liberation struggle, and helped it maintain its credentials with Western governments. At the same time it avoided the situation in Mozambique by insisting on administrative regulation of international NGOs.

By contrast, Botswana's economic growth has generated sufficient government revenues to finance universal welfare provision, allowing the government to accept the aid of Northern NGOs on its own terms. These NGOs have tended to work through government agencies rather than independently or in partnership with indigenous NGOs, which have hardly been in evidence until very recently. Despite their reduced role compared to Zimbabwe, local branches of international NGOs have dominated the Botswana NGO scene, with some of their expatriate staff also taking leading roles in local or national NGOs.

Nigeria is different again. Its authoritarian politics have not been conducive to NGO activities, whether indigenous or international, and its oil wealth has potentially provided the resources to resolve its own problems. Thus the activities of Northern-based development NGOs have been extremely limited.

Third, at the national level there are public interest research or campaigning organisations. Some, relatively few in number, are Western-style conservationist organisations. NGOs of this type with a role in the case studies in this book include the Okavango Wild Life Society (OWLS) based in South Africa, the Kalahari Conservation Society in Botswana (Chapters 6 and 7) and the Nigerian Conservation Foundation (Chapter 8).

These NGOs are analogous to conservation societies in Northern countries in that their activities include voluntary efforts in support of wildlife sanctuaries and so forth. However, in African countries with large settler populations such as Zimbabwe they are also seen as representing commercial settler interests, including hunting and safari operations as well as commercial farming. Although this makes it hard for such NGOs to be seen as disinterested campaigners on behalf of, say, wildlife conservation, it also means that in certain areas they have economic importance. Providing they keep a distance away from issues with a race element, these NGOs may be well placed to undertake issue-based campaigning.

In Botswana conservationist NGOs have tended to be expatriate-led rather than settler-led. This may mean they are less identified with particular economic interests. However, it also means it may be possible to dismiss their concerns as 'a white man's problem'. In fact, to quite a large extent the whole NGO movement in Botswana has until recently been identified with the expatriate community.

However, with the growth of an urban salaried middle class there is room for indigenous NGOs with some of the characteristics of environmental pressure groups in Europe or the USA. Both Zimbabwe and Botswana have relatively new environmental NGOs financed partly by sponsorship from commercial sources and with interests in practical projects such as waste recycling and the promotion of environmental awareness. The Matabeleland Zambezi Water Project (MZWP), set up by a coalition of industrial, commercial and political interests in Bulawayo to press for a pipeline from the Zambezi (Chapter 5), is best conceived as of this type. The Botswana Society is a national civic organisation which raises matters of public concern, including environmental issues. ZERO (originally the Zimbabwe Environmental Research Organization), based in Harare, is a 'regional network of environmental experts' with a transnational character covering the southern African region. The Nigerian Environmental Study/Action Team (NEST) draws its membership mainly from Nigerian universities. Both ZERO and NEST were involved in the International Negotiating Committee for a Convention to Combat Desertification (INCD) (Chapter 3). It is arguable that some of these NGOs should be regarded as membership organisations, existing to promote partial interests. I prefer to accept their own representations of themselves as public interest organisations.

The fourth type are indigenous, national (and provincial) service-providing NGOs, mostly concerned with welfare and rural development. Some of these aim at particular constituencies, such as women or the rural poor, while others adopt a participatory style and attempt to build up local membership. Many international NGOs have moved from managing projects themselves to working through partnerships with NGOs of this type and the next. In Nigeria, for example, despite the lack of an established NGO culture, there has been a rapid increase in NGOs since the late 1980s when the preparation for the 1992 United Nations Conference on Environment and Development increased the prominence of NGOs and highlighted their potential contribution to sustainable development. This has also been a trend elsewhere. Thus some such NGOs tend to be artificial creations of the need for local partners for international NGOs. Nevertheless, they are essentially based on trust, charity and private initiative aimed at social and development goals.

It is often argued that NGOs have distinctive features that give them comparative advantage with respect to government agencies, particularly in areas such as responsiveness, flexibility, experimentation and risk-taking (see e.g. Fowler 1988; Fisher 1993). In Botswana, indigenous NGOs are still relatively few in number, partly because of the tendency for international NGOs to work directly with government agencies, and partly because of cooption by state agencies. They do, however, include examples working in new areas of development pioneered by them, such as Thusano Lefatsheng in the promotion of 'veld products' (Chapter 6), Permaculture Trust Botswana, and others. This seems to confirm the idea of NGOs as having special capacities for innovation. Vivian (1994) notes that in Zimbabwe NGOs are expected to fulfil a qualitatively different role in rural development from that of government agencies, and in particular to be flexible and innovative. However, she reports that in practice, although there are considerable numbers of indigenous development NGOs in Zimbabwe, their 'reach' is 'both quantitatively and qualitatively limited', and '[t]he majority see themselves as "filling gaps": as extending the services normally provided by government so that more people can benefit from them' (Vivian 1994: 183, 185).

Some Zimbabwean NGOs are innovative both in technical and in institutional terms. The promotion of CAMPFIRE schemes by a consortium including the Zimbabwe Trust is one prominent example. However, there is clearly some difference in this respect between the different countries. Despite their limited reach, Zimbabwe has many indigenous NGOs working in partnership with international NGOs to provide services similar to those of government agencies. Only a few of these are developing new ideas. In Nigeria, similarly, NGOs tend to be driven not by ideals so much as by immediate needs and practical considerations. In Botswana, with state provision closer to universal, the few NGOs tend to be more distinctive in character.

Fifth, there are membership organisations existing to further their members' mutual interests. The term 'grassroots organisation' is often used to refer to local organisations of this type. Local mutual benefit NGOs may also combine into regional or national federations, associations or unions, such as the Organization of Rural Associations for Progress (ORAP) in Zimbabwe (Chapter 4). In principle there is a clear distinction between this type of NGO and those of the public-benefit type, even if the latter adopt a fully participative style of working. In practice the differences may not be so clearcut. A public benefit NGO may be set up from the top with the intention of attracting local members to whom the staff should be accountable, as in the case of the Manicaland Development Association, also in Zimbabwe. NGOs may well end up working both for member benefit and for general public benefit, particularly in environmental matters where local member interests may well coincide with protection of the environment.

One important type of member-based NGO is that based on a local environmental campaign. An example from Botswana is the Tshomorelo Okavango Conservation Trust (TOCT), featured in Chapter 7 for its opposition to the Southern Okavango Integrated Water Development Project. Other parts of Botswana also have potentially oppositional community-based environment groups. Where local political activity through other channels is suppressed, NGOs may be established to allow the political demands of a particular community or ethnic group to be expressed. In Nigeria the Movement for the Survival of the Ogoni People (MOSOP) did this via an environmental campaign (Chapter 9).

A framework for analysing NGO influence

Thus NGOs vary hugely and a great deal is expected of them. In this book we assess these expectations and how NGOs' role should be conceptualised by exploring under what circumstances NGOs are able to achieve policy influence. Each of Chapters 3 to 9 examines a case where they attempted to do so. Here a framework is provided by breaking the question down into three subsidiary questions, focusing first on the political context, then on the policies, and finally on the NGOs themselves:

A What is it about different political contexts which means NGOs can exert more or less policy influence in them?
B What is it about certain environmental policies which makes them more or less amenable to influence by NGOs?
C What is it about particular NGOs and their activities which makes it more or less likely that they will succeed in influencing environmental policies?

The following sections develop this framework, especially in relation to Africa. For each of the three questions a number of broad areas is suggested in which to seek explanatory factors. Table 1.2 (see p. 18) summarises the framework as a whole.

The political context: aspects of democracy

A major question for the research on these case studies has been how NGO campaigning is affected by domestic political factors. Of particular interest is whether NGO influence is enhanced in countries with a democratic system of government. We sited our case studies in three countries with considerably different political contexts. Botswana is generally regarded as one of Africa's few successful multi-party democracies. There has effectively been a one-party system in Zimbabwe since its independence in 1980. Nigeria has been ruled by military governments for twenty-nine of the thirty-nine years since its independence in 1960.

However, it would be an over-simplification to characterise Zimbabwe and Nigeria as 'authoritarian' and Botswana as 'democratic'. In fact there are democratic and undemocratic elements in the politics of all three countries. These are worth examining briefly before continuing with the argument.

Despite their differing reputations, Botswana and Zimbabwe have both held regular multi-party elections since their respective independence dates and in both cases the same party has won every time. Botswana is credited with open, democratic institutions. However, there has been little spontaneous development of NGOs or other civil institutions, and those which have emerged have tended to be co-opted by government, leaving civil society relatively weak (Molutsi and Holm 1990). The democratic façade conceals a small elite that has ruled the country since independence, with senior civil servants and politicians interchanging positions in a manner rare in mature democracies, but with a successful orientation towards national economic development which has been characterised as 'paternalistic developmentalism' (Charlton 1991).

In Zimbabwe white settlers played a major historical role. From the late nineteenth century land was alienated from indigenous Africans. The successful liberation struggle of the 1970s gave rise to profound changes in the political culture of the black majority, setting up far-reaching expectations, for example for land redistribution. Independence was gained in 1980 under the Lancaster House Agreement, which provided for multi-party elections and guaranteed tenure for commercial farmers for a minimum of ten years. The Zimbabwe African National Union (ZANU) won a sweeping election victory and has been in power ever since, in many cases maintaining authoritarian state institutions formed under colonialism. In the mid-1980s armed 'dissidents' from the opposition Zimbabwe African People's Union (ZAPU) were suppressed after a near civil war in Matabeleland and their leaders co-opted into the ruling party, which was renamed the Zimbabwe African National Union – Patriotic Front (ZANU-PF).

The ten-year protection clauses negotiated in 1980 have long expired. State (effectively party) control of institutions has relaxed partly because the Zimbabwean state lacks the capacity to deliver the government's policy undertakings on resettlement, or on rural development, wildlife conservation, education and welfare. The government tends to allow independent actions by donors and their local NGO partners to fill the gaps in state provision and bring in foreign exchange. Although ZANU-PF publicly favoured establishing a socialist one-party state, the government now embraces the principle of multi-party elections. ZANU-PF encounters increasing opposition particularly in urban centres and from organised labour, yet continues to dominate the electoral process.

Like Botswana, Nigeria inherited a British style of liberal democratic constitution. However, political parties have generally been regionally based and used as vehicles for personal economic advancement by their leaders. Nigeria is a federation of states, each of which has a different ethnic and religious composition, with potential for internal conflict epitomised by the 'Biafran' war of the late 1960s. Military coups have been justified by reference to the need to enforce national unity or stamp out corruption. Successive military governments have promised or even embarked on a transition back to democracy, but on several occasions the process has been aborted or the subsequent civilian government has remained dependent on military support.

Despite the lack of any countervailing force to restrain the military, Nigeria's vibrant and literate population remains irrepressible. Although

Nigerian radio and television are government-run, most newspapers and magazines are independent, lively and often outspoken. Despite some instances of manipulation by the military, in general the Nigerian judicial system has maintained its autonomy (Lewis 1994). There is also a marked gap between the traditional institutions of society at the grassroots level and the superimposed state structures (Vaughan 1995). Lack of co-ordination between federal, state and local government institutions, inadequate funding of development projects and widespread corruption in the newly created bureaucracies mean that at local level, particularly in rural areas, traditional institutions and community-based organisations play significant roles. During the extended period of military rule of the 1990s many international aid organisations (including the UK's then Overseas Development Administration) channelled financial support directly through local civilian authorities and NGOs rather than via the military government.

Even these brief sketches illustrate that the differences between the three political contexts cannot be fully characterised as 'more' or 'less' democracy. Hence we have unpacked the concept of democracy and developed a research framework which can test the circumstances under which NGOs may be influential, even when the received wisdom might classify the countries they work in as 'authoritarian'. Rather than seeing the political context in terms of 'more' or 'less' democracy, we describe it in terms of five aspects of democracy which may occur in various combinations. The first two form the basis of many definitions of democracy (e.g. Lafferty and Meadowcroft 1996: 3–4):

1 The rulers are accountable to the ruled through periodic multi-party elections based on universal suffrage.
2 There are guarantees in law of political and human rights.

In addition we suggest three other features of a fully democratic society, the last of which in particular is often invoked by proponents of one-party or 'no-party' forms of African democracy:

3 There is a plurality of competing sources of power within and outside the state. In other words, and in contradistinction to an authoritarian political system with a relatively homogeneous state apparatus, there may be disagreement between ministries, local councils and other actors.
4 There is a limited reach of the state, that is, there are areas of political and economic life over which the state has ceded control to other actors.
5 There is a political culture of participation so that people throughout society are able to partake in decision-making processes.

Each country features some of these aspects of democracy more than others. We use our case study material to explore the idea that different combinations favour different types of NGO campaign strategy.

Environmental policies in an African context

Environmental policies, and how they change in response to various pressures including NGOs, relate to broader questions of sustainability and sustainable development. However, these remain contested concepts, involving real and often strongly conflicting interests. They require mediation: 'a process of "coming between" different interests with a view to finding a way forward from what is, or is in danger of becoming, a cul de sac of conflict or inertia' (Blauert and Zadek 1998: 10).

Sustainability can mean sustaining the status quo, under which many at the grassroots lack basic needs. Alternatively, the concept can be applied to livelihoods, to resource utilisation, to institutions, even to economic growth. Typically those with the power to formulate and implement policy, in other words those who might be 'targeted' by NGOs, conceive sustainability differently from those affected by the policies, whose interests may to a greater or lesser extent be represented by the NGOs.

The environmental policies on which our research focused relate mainly to land degradation and sustainable resource management. The general characteristics of land degradation as an international issue and the main lines of debate about it are discussed in Chapter 2. Chapter 3 concerns the role of NGOs in the negotiation and implementation of the international Convention to Combat Desertification. Chapters 4 to 9 examine some of the many different specific policy issues that arise within national contexts. These policies each have particular characteristics which may make them more, or less, amenable to NGO influence. We suggest three areas where explanations may be sought.

1 The nature of the issue

Most often NGOs relate to environmental issues by seeking to protect a feature of the environment from over-exploitation or degradation resulting from the policies of state or private actors. However, there are generally complex ramifications to any issue which go beyond a simple opposition of development or exploitation versus environmental protection, and which affect the likelihood of success for NGOs.

Most obviously, issues which directly affect the interests of powerful actors are less likely to be amenable to immediate NGO influence. For example, attempting to change state policy may be much harder if the issue concerns matters of central economic or political importance. Thus in Botswana any proposal which might adversely affect the diamond or beef industries, or the government's ability to maintain popularity through rural welfare schemes, would be difficult to advance.

Another important question is how an issue relates to questions of national sovereignty and global or local commons. Often there are conflicting custodial claims to an environmental resource. Several of our cases relate to competing uses for water resources, with a government's claimed right to develop a resource to meet national needs clashing either with community claims to a local common or with claims for the global importance of a resource, for example its contribution to global biodiversity or climate regulation.

A related question concerns possible differences between public and private interests. For example, issues of land tenure underlie many of the problems of land degradation in Africa. Private interests often assert that traditional forms of land tenure involving local commons lead to over-exploitation since no individual has overall responsibility. Private ownership is said to be required before anyone will invest in measures to enhance or protect environmental resources. However, these arguments ignore the traditional institutions which regulate the use of 'common' land, and the fact that private ownership is no guarantee of good land management. Also it is probably security of tenure rather than its particular form which is a prerequisite for investment in land. Arguably, whatever the form of land tenure it is crucial that the maintenance of the environment be seen as a matter of common or public interest rather than a private matter for a landholder.

Finally, many environmental issues concern intergenerational equity in a context of scientific and social uncertainty. In other words, maintaining a balance between the needs of current and future generations is made difficult by not knowing the precise outcomes likely to result from proposed actions. The precautionary principle would dictate either not acting unless the possibility of adverse effects can be completely ruled out, or else acting only if precautions can be taken against all possible outcomes. However, where proposed development aims to meet clear and immediate needs, it may be difficult for an NGO to advocate caution and hence oppose that development without definite evidence of likely harm.

2 *The character of the organisations targeted*

Some organisations with policies affecting the environment are more amenable to influence than others, depending on to whom they are accountable, the mechanisms for that accountability, and the political and economic interests at stake. Thus transnational corporations, as private firms, are accountable to their shareholders, though they also require good relations with governments and communities in order to operate effectively. Even if TNCs are not directly affected by NGO activity, they may be open to indirect influence via consumer boycotts or government regulation or the threat of them, and may avoid confrontation if it would damage their standing in the market.

Multilateral agencies differ considerably in their readiness to accept outside influence. Most of the UN specialised agencies allow some access to NGOs. The World Bank has become more open in recent years, and several coalitions of NGOs have campaigned to influence it (Fox and Brown 1998). However, other multilateral agencies, such as the World Trade Organization, have proved much more impervious to outside influence.

Most of the cases in this book concern attempts to influence states or government agencies. The question of the degree or form of democracy has been considered above. Another important characteristic of any state is its capacity in respect of development. 'Developmentalist' states (see e.g. Leftwich 1994) may be amenable to some forms of NGO influence, but those states with less capacity to undertake independent development programmes may need NGO activities to bring in resources for development, so that NGOs trying to affect environmental policies start from a stronger position.

3 The different stages of policy development

We conceive policy, following Grindle and Thomas (1991), as a continuous process with three distinguishable phases: agenda-setting, policy choices and implementation processes. As noted above, Willetts argues that at least within UN negotiations NGOs tend to 'dominate agenda-setting'; he also suggests that they are a 'major influence on policy implementation', but that when it comes to the middle part of the policy process, 'governments and diplomats ... tend to give the highest priority to policy formulation and are less subject to NGO influence in this process' (1996: 46).

At national level a similar view of the role of NGOs vis-à-vis governments would see them on the one hand as a kind of interest group feeding information into the policy process and highlighting issues of concern. On the other hand they have a presence on the ground and thus are also important in policy implementation, in many cases being given a direct role. However, the decision-making itself is arguably the legitimate concern only of public authorities. In practice policy choices may be influenced by NGOs, but such influence may be more difficult or require different strategies from influence over agenda-setting or implementation. This may apply to attempts at influencing other policy agencies, not just governments.

C. NGOs' ability to exert influence

We turn now to the third question in our framework: what is it about NGOs themselves and what they do which explains their degree of success at influencing environmental policy? Here we should note that it is not just 'environmental' organisations established for advocacy purposes which may influence environmental policies. A great variety of types of NGO, like those discussed above, may be involved. NGOs of different types have different histories and concerns. Not surprisingly, we are likely to find that they attempt to influence environmental policies in very different ways. Thus we should look for explanations which reflect these differences. Four such areas of explanation are suggested here.

1 Legitimacy and capacity of NGOs

It is unlikely that NGOs will be able to influence policy unless they are regarded as legitimate representatives of the viewpoints they put forward. They will also be unable to influence policy without the capacity to present those viewpoints to good effect. There are several possible sources of legitimacy. A broad-based membership among a particular population may allow an NGO to speak authoritatively on behalf of that population. A development or service-providing NGO may claim to represent the interests of those it serves (though this may not automatically be accepted) or may claim the right to promote policy options on account of the expertise gained from grassroots development work. A campaigning environmental organisation will claim to represent the public interest rather than a particular group, and again base its legitimacy on expertise and reputation.

Expertise is also an important aspect of capacity, along with technical skills appropriate to the particular issue, presentation, communication and political skills. Finally, the way an NGO is managed will contribute significantly to its capacity to capitalise on opportunities for influence.

2 NGO strategies and their appropriateness

Several writers have discussed the range of strategies or advocacy approaches open to NGOs attempting to influence policy. Drawing on the ideas of Moyo (1991), Clark (1991) and others, I have previously developed a schema of four ways in which NGOs try to achieve influence, namely the 'four Cs' (Thomas 1996: 44):

1. collaboration (including reform and entryism);
2. confrontation (or opposition, including passive resistance);
3. complementary activities (projects or programmes carried out independently of government or other decision makers, whose success can force a change of policy to accommodate them);
4. consciousness-raising (indirect, generalised campaigning).

Different strategies based on different combinations of these 'four Cs' may be more or less appropriate depending on the presence or absence of the features of democracy explained above. However, it should be noted that NGOs generally cannot freely choose their advocacy strategy. In particular, collaboration with government or other target institutions presupposes a degree of common ground while complementary activities may not always be permitted. Conversely, confrontation may be unlikely to achieve immediate results, but may be the only option if there is complete disagreement between the NGO and the target institution. In some cases campaigning may be subject to repression, but the NGOs concerned may feel morally obliged to continue to campaign.

3 Relations between NGOs

Most often more than one NGO is involved in any given issue, and their interrelationships will affect the influence they can bring to bear. For example, if they represent interests which diverge with respect to how they are affected by the issue at hand, there may be no discernible 'NGO position' and whatever influence they have may be cancelled out. Alternatively, where possible, NGOs may try to increase the resources available to put pressure on a 'target' organisation by working with others. The degree of intensity of collaboration can vary considerably. Eccleston (1996) distinguishes between informal networking, networks, coalitions and alliances, in increasing order of potential for mobilising resources and widening the political space for a campaign. However, increased cooperation also increases costs in terms of the time and resources required to maintain coordination.

Networking can be a way of internationalising an issue by involving NGOs from other countries. NGOs from African countries may effectively by-pass their own governments by dealing directly with international NGOs, and indeed NGOs' frame of reference is increasingly oriented towards the global rather than the state. However, international networking may have negative as well as positive effects, particularly as there is a risk of provoking a backlash, since such actions may be interpreted as effectively challenging the sovereignty of the state.

4 How an issue can be framed

When an issue comes to the fore it is not well-defined beforehand. Exactly what the issue is and how it is perceived will depend on how it is framed by those to whom it is salient. This is always subject to contestation. An NGO will have much more chance of influencing policy choices on an issue if its definition is accepted by other actors. For example, it may be important to frame an issue as an international rather than purely national one, or in terms of common interests rather than a dispute between different private interests.

Even if an NGO initially fails to influence a particular policy, it may succeed in redefining an issue so that over time a new type of public need is recognised.

Table 1.2 Summary of research framework

A	*What is it about different political contexts which means NGOs can exert more or less policy influence in them?*	
	1	free and fair multi-party elections
	2	political and human rights
	3	plurality of sources of power within and outside the state
	4	limited reach of the state
	5	a culture of political participation
B	*What is it about certain environmental policies which makes them more or less amenable to influence by NGOs?*	
	1	the nature of the specific issue with respect to sustainable development and to the interests involved;
	2	the character of the organisations being targeted by the NGOs;
	3	the different stages of policy development.
C	*What is it about particular NGOs and their activities which makes it more or less likely that they will succeed in influencing environmental policies?*	
	1	the legitimacy and capacity, including expertise and skills, of the NGOs themselves;
	2	the appropriateness of the strategies adopted by NGOs (the 'four Cs');
	3	the relations between NGOs, including their degree of unity, use of networking, and the involvement of international agencies of different types;
	4	the way NGOs frame the issue, including whether they are able to portray it as an international issue.

Contributions to this volume

In Chapter 2 David Humphreys and Susan Carr lay out some of the issues in the debate about land degradation and desertification, which is the context for the case studies explored in the following seven chapters. It considers various definitions of land degradation and gives a brief history of the international debate on the issues, including theories about the different ways they have been socially constructed and alternative ideas on their causes and effects. It concludes by considering the issues in terms of competing claims to resources between the global, national and local levels.

Chapter 3 is the first case study, on the role of NGOs at the International Negotiating Committee for a Convention to Combat Desertification (INCD) between 1993 and 1996. Four of the chapter's co-authors attended the INCD negotiations with observer status, two of them representing African NGOs, Enoch Okpara as a member of the Nigerian Environmental Study/Action Team (NEST) and Roger Mpande on behalf of the Zimbabwe Environmental Research Organization (ZERO). The convention, although international, gives priority to action in Africa, and consequently African governments and African NGOs played a prominent part in the debate. The influence that NGOs had on the agenda setting and policy choice stages of the negotiations process is examined, as is the influence of the negotiation process on the organisation and tactics of the NGOs themselves. NGOs have secured a major and innovatory role in the implementation of the convention's provisions, and the chapter concludes by appraising the potential effects of this development.

Chapters 4 to 9 present cases of NGO influence (or failure to influence) in Zimbabwe (Chapters 4 and 5), Botswana (Chapters 6 and 7) and Nigeria (Chapters 8 and 9). In Chapter 4 Derek Gunby of PlanAfric, a Zimbabwean consultancy based in Bulawayo, examines the role of NGOs in promoting the use of small grains in Zimbabwe as an alternative to maize, the dominant staple crop in the country. Small grains offer a means of increased food security for small farmers in semi-arid regions while minimising pressure on the fragile environment. Despite the entrenched dominance of maize, the campaign has had some, albeit limited, success through NGO activity at the grassroots level. In Chapter 5, Gunby, Mpande and Thomas analyse the campaign for water from the Zambezi river. The availability of water and the ways in which it is managed is central to Zimbabwe's development and environmental management. Water shortages are particularly acute in Matabeleland and Bulawayo, prompting a local NGO to initiate a campaign that has influenced government policy.

In Chapter 6, Onalenna Selolwane of the University of Botswana discusses the role of NGOs in opposing the fencing policy introduced in 1991 by the Botswana government to help promote the country's cattle industry. This chapter examines the failure of Botswana NGOs to influence policy change with respect to rangelands fencing, despite their evidence that fencing has contributed to increased land degradation, and critically assesses the weaknesses of the NGOs involved. In Chapter 7, Thomas, Selolwane and Humphreys analyse the campaign against the proposed development of the Okavango Delta in 1990 and the subsequent policy change by the Botswana government. Central to the campaign was the involvement of international NGOs and their relationship with local and national NGOs in Botswana.

In Chapter 8, Susan Carr and Obasi Ogbonnaya examine the campaign to preserve the natural flooding of wetlands in semi-arid northern Nigeria, an area that is internationally important for migrating birds as well as the source of livelihood for thousands of smallholders, fisherfolk and graziers. The tactics and strategies employed in the campaign by a prominent Nigerian conservation NGO and a related local community development project are analysed, as are the roles of international conservation and research organisations. The influence of local, national and international forces on the decision to shelve the construction of further large dams is discussed. In Chapter 9 Carr, with Oronto Douglas and Uche Onyeagucha of the Nigerian NGO Environmental Rights Action, examines the role of NGOs in promoting human and environmental rights in the Niger Delta.

The chapter charts the struggle of the Ogoni people, a minority ethnic group living in the delta, for self-determination, equitable access to the benefits from the development of oil resources, and amelioration of the environmental damage caused by that development. It analyses the tactics by which the Ogoni people and the NGOs associated with them have attracted international attention to their cause and gained support in Europe and North America. It examines the strategies used to influence the Nigerian government and the Shell oil company.

The final chapter returns to the framework developed here. It considers the role of NGOs in a global context and examines the lessons for NGO campaigners and policy analysts from the case studies in Chapters 3 to 9. As well as analysing how NGO influence can be explained in terms of the three elements of the framework, namely the political context, the environmental policies and the NGOs themselves, the chapter considers the relationship between direct policy influence and longer-term, indirect effects on how issues are defined globally, and the way that policy processes can influence NGOs as much as NGOs influence the policies. Finally, the chapter suggests using the concept of public action in order to conceptualise NGOs and their role in designing and implementing policies for sustainable development in the new millennium.

References

Annis, S. (1988) 'Can small-scale development be large-scale policy?', in S. Annis and P. Hakim (eds) *Direct to the Poor: Grassroots development in Latin America*, London: Lynne Rienner.

Blauert, J. and Zadek, S. (1998) *Mediating Sustainability: Growing Policy from the Grassroots*, West Hartford, Conn.: Kumarian Press.

Charlton, R. (1991) 'Bureaucrats and politicians in Botswana's policy-making process: a re-interpretation', *Journal of Commonwealth and Comparative Politics* XXIX, 3: 265–82.

Clark, J. (1991) *Democratizing Development*, London: Earthscan.

Eccleston, B. (1996) 'Does North–South collaboration enhance NGO influence on deforestation policies in Malaysia and Indonesia?', *Journal of Commonwealth and Comparative Politics* XXXIV, 1: 66–89.

Edwards, M. and Hulme, D. (1992) *Making a Difference: NGOs and Development in a Changing World*, London: Earthscan.

Edwards, M. and Hulme, D. (1995) 'NGO performance and accountability in the post-cold war world', *Journal of International Development* 7, 6: 849–56.

Fisher, J. (1993) *The Road from Rio: Sustainable Development and the Non-governmental Movement in the Third World*, Westport, Conn.: Prager.

Fisher, J. (1998) *Nongovernments: NGOs and the Political Development of the Third World*, West Hartford, Conn.: Kumarian Press.

Fowler, A. (1988) 'Non-governmental organizations in Africa: achieving comparative advantage in relief and micro-development', *IDS Discussion Paper* 249, Brighton: Institute for Development Studies.

Fowler, A. (1993) 'Non-governmental organizations as agents of democratization: an African perspective', *Journal of International Development* 5, 3: 325–9.

Fox, J. and Brown, L.D. (eds) (1998) *The Struggle for Accountability: The World Bank, NGOs and Grassroots Movements*, Cambridge, Mass.: MIT Press.

Ghai, D. and Vivian, J. (eds) (1992) *Grassroots Environmental Action: People's Participation in Sustainable Development*, London: Routledge.

Grindle, M. and Thomas, J. (1991) *Public Choices and Policy Change: The Political Economy of Reform in Developing Countries*, Baltimore: Johns Hopkins University Press.

Handy, C. (1988) *Understanding Voluntary Organizations*, Harmondsworth: Penguin.

Hanlon, J. (1991) *Mozambique: Who Calls the Shots?*, London: James Currey.

Korten, D. (1987) 'Third generation strategies: a key to people-centred development', *World Development* 15 (Supplement): 145–60.

Lafferty, W.M. and Meadowcroft, J. (1996) 'Democracy and the environment: congruence and conflict – preliminary reflections', in W.M. Lafferty and J. Meadowcroft (eds) *Democracy and the Environment*, Cheltenham: Edward Elgar.

Landim, L. (1993) 'Brazilian crossroads: people's groups, walls and bridges', in P. Wignaraja (ed.) *New Social Movements in the South: Empowering the People*, New Delhi: Vistaar.

Leftwich, A. (1994) 'Governance, the state and the politics of development', *Development and Change* 25, 2: 363–86.

Lewis, P. (1994) 'Endgame in Nigeria? The politics of a failed democratic transition', *African Affairs* 93: 323–40.

Molutsi, P. and Holm, J. (1990) 'Developing democracy when civil society is weak', *African Affairs* 89, 356: 323–40.

Morphet, S. (1996) 'NGOs and the environment', in P. Willetts (ed.) *The Conscience of the World: The Influence of Non-governmental Organisations on the UN System*, London: Hurst & Co.

Moyo, S. (1991) 'NGO advocacy in Zimbabwe: systematising an old function or inventing a new role?', paper to workshop on *NGO and Development Advocacy*, organised by IRED and 'NGO/PVP Initiatives Project', Harare: ZERO (mimeo).

OECD (1998) *Development Co-operation: Efforts and Policies of Members of the Development Assistance Committee* (1997 Report), Paris: OECD.

Potter, D. and Taylor, A. (1996) 'Introduction', *Journal of Commonwealth and Comparative Politics* XXXIV, 1: 1–8.

Princen, T. and Finger, M. (1994) *Environmental NGOs in World Politics: Linking the Global and the Local*, London: Routledge.

Riddell, R. (1990) 'Judging success: evaluating NGO approaches to alleviating poverty in developing countries', *ODI Working Paper* 37, London: Overseas Development Institute.

Riddell, R. and Robinson, M. (1995) *Non-governmental Organizations and Rural Poverty Alleviation*, Oxford: Clarendon Press and London: Overseas Development Institute.

ROAPE (1992) 'Democracy, civil society and NGOs' (Editorial Introduction), *Review of African Political Economy* 55: 3–7.

Sloep, P.B. and van Dam-Mieras, M.C.E. (1995) 'Science on environmental problems', in P. Glasbergen and A. Blowers (eds), *Environmental Policy in an International Context: Perspectives on Environmental Problems*, London: Arnold.

Thomas, A. (1992) 'Non-governmental organizations and the limits to empowerment', in M. Wuyts, M. Mackintosh and T. Hewitt, (eds) *Development Policy and Public Action*, Oxford: Oxford University Press.

Thomas, A. (1995) 'Does democracy matter? – Pointers from a comparison of NGOs' influence on environmental policies in Zimbabwe and Botswana', *DPP Working Paper 31/ GECOU Working Paper* 4, Milton Keynes: GECOU, The Open University.

Thomas, A. (1996) 'NGO advocacy, democracy and policy development: some examples relating to environmental policies in Zimbabwe and Botswana', *Journal of Commonwealth and Comparative Politics* XXXIV, 1: 38–65.

UNDP (1993) *Human Development Report*, New York: United Nations Development Programme.

Union of International Associations (1996) *Yearbook of International Organization*, 33rd edn, Munchen: K.G. Saur.

Uphoff, N. (1993) 'Grassroots organizations and NGOs in rural development: opportunities with diminishing states and expanding markets', *World Development* 21, 4: 607–22.

Van Rooy, A. (ed.) (1998) *Civil Society and the Aid Industry*, London: Earthscan.

Vaughan, O. (1995) 'Assessing grassroots politics and community development in Nigeria', *African Affairs* 94: 501–18.

Vivian, J. (1994) 'NGOs and sustainable development in Zimbabwe: no magic bullets', *Development and Change* 25, 1: 167–93.

Wapner, P. (1995) 'Politics beyond the state: environmental activism and world civic politics', *World Politics* 47: 311–40.

Wignaraja, P. (ed.) (1993) *New Social Movements in the South: Empowering the People*, New Delhi: Vistaar.

Willetts, P. (1990) 'Transactions, networks and systems', in A. Groom and P. Taylor (eds) *Frameworks for International Co-operation*, London: Pinter.

Willetts, P. (ed.) (1996) *The Conscience of the World: The Influence of Non-governmental Organisations on the UN System*, London: Hurst & Co.

2 Land degradation and sustainable resource management in an African context

David Humphreys and Susan Carr

This chapter explores the issues of land degradation and sustainable resource management, which form the context for any environmental policy work by NGOs in sub-Saharan Africa. Various definitions of land degradation and desertification are considered and a brief history of the international desertification debate is provided, including responses within the United Nations system. Attention is given to the various ways the issues of land degradation and desertification have been socially constructed. The causes and effects of degradation are discussed, although it is emphasised that scientific uncertainty remains with respect both to climatic and human-induced causes. Competing claims to land ownership between the state and local levels in Africa are examined. Consideration is given to the works of African scholars.

Defining the issues: land degradation and desertification

Drylands cover about 30 per cent of the earth's terrestrial surface and are inhabited by about 900 million people. Degradation of dryland areas occurs in the Americas and in Asia, but is particularly acute in Africa where it is one of the most serious problems the continent faces. Beginning with the Sahel drought of 1968, land degradation and desertification have attracted international attention, culminating in the conclusion of negotiations for the United Nations Convention to Combat Desertification in 1994. The convention, which entered into legal effect on 29 December 1996 following the deposit of the fiftieth ratification by Chad, assigns a key role to non-governmental organisations (NGOs) in implementing projects to combat desertification and to manage land resources sustainably. Over recent years NGOs have increased in importance both in the management of development projects and in national and international lobbying and advocacy. The convention recognises the importance of NGOs while also opening up potential new spheres of NGO activity at the national and local levels for issues beyond the remit of the convention.

The United Nations Convention to Combat Desertification defines, for the purposes of the convention, land degradation as 'reduction or loss in arid, semi-arid and dry sub-humid areas, of the biological or economic productivity and complexity of rainfed cropland, irrigated cropland, or range, pasture, forest and woodlands resulting from land uses or from a process or combination of processes, including processes arising from human activities and habitation patterns' (United Nations 1994: Article 1 (f)). The emphasis in this definition on arid, semi-arid and dry sub-humid areas is explained by the fact that the convention's principal focus is desertification in these areas. The convention is not concerned with land degradation in other areas.

Land degradation involves a loss of soil fertility, broadly defined to encompass 'the range of solid chemical, physical and biological factors which affect the productive potential of the land' (Scoones and Toulmin 1999: 22). It has physical, biological, economic and social aspects, although the concept of degradation only has meaning in economic and social terms. Reduction in the fertility of soil may be offset in the short term by increases in inputs of labour, nutrients, or fuel and machinery. If the degradation is not too severe, some remediation may be afforded by natural processes, but degradation almost always gives rise to economic or social costs in the medium to long term.

The causes and forms of land degradation vary widely. In its most extreme form, land can be irretrievably lost through urbanisation (the conversion of the natural environment to the built environment) or through inundation by dam construction. Campaigns by NGOs against dam construction in Botswana and Nigeria are covered in Chapters 7 and 8. Other forms of degradation may involve soil loss through erosion, or changes in the biophysical nature of the soil itself. Erosion can take many forms, but is greatly increased by the use of soil management practices inappropriate to a given climate and soil type (Morgan 1986). Less overt changes in soil structure can reduce the potential for crop growth and may be brought about by compaction of the soil through inappropriate use of machinery. These effects may be exacerbated by loss of appropriate forms of organic matter in the soil. These losses occur through oxidation, failure to return crop residues, or through changes in the soil microflora and fauna induced by injudicious use of pesticides and cultivation techniques (Edwards and Lofty 1977, 1982). In extreme situations, this can lead to the formation of a severely indurated surface pan on which normal plant growth is impossible. High level usage of certain synthesised fertilisers can reduce soil pH, causing problems for some crop species. Salinisation, which can severely restrict the growth of most crops, occurs particularly when the balance between the rate of irrigation and evaporation is wrongly chosen. The human role in land degradation is critical. As Blaikie and Brookfield note:

> Leaching, compaction and erosion of the soil, changes in plant cover and hydrology regime, changes in soil and water chemistry all take place naturally ... Yet human interference has modified and usually

> accelerated all these processes and has created the conditions under which new sets of processes, previously absent or insignificant, come into play.
>
> (Blaikie and Brookfield 1987: 3)

Expressed simply, desertification is land degradation in dry areas (Grainger 1990). Agenda 21, adopted at the United Nations Conference on Environment and Development (UNCED), defined desertification as 'land degradation in arid, semi-arid and dry sub-humid areas resulting from various factors, including climatic variations and human activities' (United Nations 1993: chapter 12). This definition also appears in the Convention to Combat Desertification (United Nations 1994). Mainguet sees desertification as 'the ultimate step of land degradation: irreversibly sterile land, meaning irreversible in human terms and within practicable economic limitations' (1994: 16). Both land degradation and desertification therefore involve a loss in the biological productivity of land, but whereas land degradation may be seen as a continuum, desertification results in an irreversible end state whereby biological productivity is permanently lost. To Hare therefore, desertification 'is the name given to the processes whereby ecosystems lose [the] capacity to revive or repair themselves' (Hare 1985, cited in Mainguet 1994: 287). Noting that the UNCED definitions did not address the concept of irreversibility, Mainguet reiterated her earlier emphasis on the irreversibility of desertification, noting that it embraces two considerations; that of time and that of economy. With respect to time, irreversibility corresponds to a period of at least one human generation. With respect to economy, desertification is irreversible as large economic inputs are necessary, with no guarantee of successful results (Mainguet 1994: 288–9).

Hence there may be said to be two principal differences between land degradation and desertification. First, desertification, unlike land degradation, is a process that occurs almost entirely in areas of aridity, semi-aridity and dry sub-humidity. Second, land degradation applies to types of degradation other than, and in addition to, arid, semi-arid and dry sub-humid areas. However, any attempt to unpack definitions of land degradation and desertification reveals how open to interpretation they are. Indeed some commentators question whether or not a problem actually exists, at least of the type and on the scale that is claimed (Morris 1995; Thomas and Middleton 1994).

When the term desertification was first used by Aubréville in 1949 it was in a different context from that in which it is generally used today. It referred to the conversion of tropical and sub-tropical forests into savannahs, involving severe soil erosion, changes in soil properties and invasion by dryland plant species (Aubréville 1949, quoted in Adams 1990). Since then, perceptions of the problem, its causes and effects have evolved and become steadily more complex, although not necessarily more certain.

Cultural and ideological influences have played an important part in shaping explanations (Adams 1990). What is land degradation to one person may not be so for another. Deforestation, often seen as a form of land degradation, may release agricultural land for the rural poor or for industry. Hence while forest clearance may constitute degradation in an ecological sense, in that it leads to physical changes to the environment, it need not necessarily constitute degradation in a social sense (Blaikie and Brookfield 1987: 6). Perceptions vary according to the values and perceived interests of the actors involved, with definitions and explanations constructed to further the objectives of particular actors. At the national level this process may often go unnoticed when the 'expertise' of national authorities is unquestioned. But in intergovernmental negotiations, the normal within-country acceptance of the views of established sources of authority and expertise no longer automatically applies. The existence of competing definitions becomes more obvious when cultural differences throw them into sharper relief. Indeed, and as Chapter 3 of this volume outlines, competing definitions were exploited by particular interests in the negotiations for the Convention to Combat Desertification, a factor that had ramifications for those NGOs seeking to influence the negotiations. NGOs had more political space within which to manoeuvre and present their own definitions and explanations; these views in turn reflected the interests of the NGOs' constituencies.

The causes of land degradation and desertification

The issue of desertification is now re-established on the global environmental agenda following a loss of international visibility after the 1977 United Nations Conference on Desertification held in Nairobi. However there remains considerable lack of clarity about the extent, nature and causes of desertification due to the complexities and scientific uncertainties involved. Unlike some environmental issues of global importance, such as climate change and ozone depletion, no authoritative transnational epistemic community has emerged that has agreed on the causes of desertification and a common policy enterprise to address the problem.[1] Hellden has argued that 'as long as the effects of desertification ... in Africa and elsewhere are not documented according to scientific standards, there is a risk that the desertification issue will become a political and developmental fiction rather than a scientific fact' (1991: 139).

Explanations offered on the causes of desertification have ranged from the local to the global and include natural, social, political and economic factors such as local mismanagement of land exacerbated by unreliable rainfall, global imbalances in trade and global climate change (Adams 1990; Grainger 1990; Mortimore 1993; Toulmin 1993). Causes vary from area to area, with the impact of the processes involved depending on the sensitivity and resilience of the soil as well as social factors. This section will consider the causes of land degradation and desertification, although it is first noted that the precise interrelationship of causal factors remains the subject of uncertainty. The section begins with an historic overview of the debate since the 1930s before briefly surveying contemporary knowledge on climatic and human-induced causes. It concludes with a neo-Marxist analysis.

Early explanations concerning local mismanagement of land originate from the period of European colonisation. For example, in the early decades of British and French colonial rule in West Africa there was concern that the limit of the Sahara Desert was moving southwards. Shifting cultivation, burning, grazing and browsing leading to removal of the natural vegetation cover were initially seen as the main contributory causes, with variable rainfall as a subsidiary factor (Stebbings 1935, quoted in Adams 1990). When subsequent research found no evidence of either desert encroachment or a permanent reduction in rainfall (Jones 1938, quoted in Adams 1990), the emphasis shifted to more general concerns about soil erosion, the main causes of which were seen as continuous cropping without nutrient replacement, over-grazing and increases in population leading to more intensive use of land that was marginal for agricultural production.

Following the Sahelian droughts of 1968–73 and 1982–4 previous perceptions of mismanagement were challenged. For example, the concept of 'overgrazing' was questioned on the basis that judgements about the land's carrying capacity for stock, which has to be judged in relation to management objectives, are inevitably subjective. A high stock density may result in poor productivity per animal but can give greater productivity per unit area. In practice, indigenous pastoral systems may be well adapted to uncertain rainfall and dryland conditions. For example, herds often include browsing animals such as camels, as well as grazers such as sheep and cattle. In dry conditions herds are moved to places where vegetation is more plentiful.

In 1991 UNEP defined desertification as 'land degradation in arid, semi-arid and dry sub-humid areas resulting mainly from adverse human impact' (UNEP 1992: vii). As noted above, the definition of desertification adopted by the UNCED and in the Convention to Combat to Desertification included both 'climatic variations and human activities'. Hulme and Kelly note that the 'different emphasis placed on climate variation in these two definitions is indicative of the disagreement that exists concerning the relative importance of the various causes of dryland degradation' (1993: 6). For example, disagreement continues on the extent to which, if at all, the Sahara is migrating southwards and whether the drought that struck in the Sahel in 1968 is a temporary or a more permanent phenomenon. While satellite evidence suggests southward retreating vegetation during the Sahel drought of 1982–4, the years from 1984–7 saw a 'generally northward advancing vegetation front' (Nelson, quoted in Forse 1989: 32).

The uncertainty and conflicting evidence on the permanence or otherwise of drought is clear from United Nations statements on the subject since the 1970s. The report of the 1977 United Nations Conference on Desertification noted that 'a review of rainfall records in the area has led to the conclusion that the Sahelian drought falls within the range of expectable climatic phenomena, something to be expected at long intervals, perhaps three or four times a century' (United Nations 1977: 6). However, by the 1990s, when preparations were under way for the UNCED, United Nations sources were claiming that there is now evidence of climatic regression in the Sahel: 'drought has become a persistent phenomenon to the extent that the northern limit of rain fed agriculture in the Sahel has moved to the South' (United Nations 1991: 11). Yet it was also noted that droughts are not the main cause of desertification, nor does desertification necessarily cause drought, although the processes leading to desertification may be accelerated by drought which, in turn, may be prolonged by desertification (United Nations 1991: 4). While there is no clear causal relationship between drought and desertification, the interaction between naturally occurring drought and unwise land use may cause desertification. As Mainguet argues, '[m]an is both the creator and the victim of land degradation but drought is the revealer and the accelerator' (Mainguet 1994: 292).

Climate change may be both a cause and a consequence of desertification, although considerable uncertainty remains on the precise relationship between climate change and desertification. In a report by the secretariat of the 1977 United Nations Conference on Desertification, Hare concluded that major climatic shifts are related to changes in global circulation and interactions between the oceans and the atmosphere. He did, however, qualify this view, noting that unwise land use during periods of drought will lead to soil erosion and gullying, thus weakening or destroying perennial vegetation which protects the soil. This, in turn, will 'introduce a new microclimate hostile to the re-establishment of vegetation and stabilization of the soil' (1977: 115). Hence even on a local scale the relationship between desertification and the climate is a circular process. Furthermore, when such changes affect larger areas 'positive feedback processes are likely to be set in motion tending to accentuate anomalies in the macro-climate' (Hare 1977: 115).

A further uncertainty in the climate change–desertification relationship concerns global warming. Hulme and Kelly note that progressive loss of trees and vegetation in desertified areas is likely to reduce the earth's carbon sink capacity. They note, however, that the contribution of desertification to the net carbon emission component of global warming 'is almost certainly less than a few per cent of the global total' (1993: 9). Present scientific knowledge is unable to provide an estimate of desertification's contribution to net emissions of other greenhouse gasses. In addition to contributing to global warming, deforestation due to, for example, the removal of wood for fuel is also problematic with respect to other features of arid zone ecology. Fallen trees and hollowed branches offer protection for many species. Rotting timber becomes a link in the food chain and a source of nutrients for desert wildlife. A good cover of trees and vegetation stabilises sand; without it sand dunes will gradually shift position, wandering and engulfing other vegetative cover in their path. Hence in some parts of Africa local communities have been encouraged to plant trees to act as windbreaks and to stabilise sand dunes (Munslow et al. 1988: 158; Leach and Mearns 1988: 154), although some observers question the efficacy of the windbreak policy, noting that active dunes rarely threaten valuable land (Forse 1989: 32). There is general agreement that more research is needed to improve scientific understanding of the relationships between climate change, deforestation and desertification.

Some analysts have concluded that there is insufficient evidence to support claims that the primary causes of desertification are human induced (Hellden 1991: 139–40; Morris 1995). However, Malthusians see population growth as a factor that leads to more *extensive* land use. One example of a Malthusian view is provided by Paul and Anne Ehrlich, who argue that in order to arrest environmental degradation population control must be a first priority (1990: 190). With respect to desertification, the Ehrlichs note the role of bad land use planning and of socio-economic conditions, before asserting that 'too many people' is one of the major causes (1990: 127). A different view is provided by Boserupians who see an increased population as catalysing technical and technological innovation thus enabling more *intensive* use of land already under cultivation (Boserup 1981).[2] A study by Mary Tiffen and others in the Machakos District of Kenya, an area with heavily degraded soil in the 1930s, supports the Boserupian rather than the Malthusian view. Population increases in Machakos have led to improved land-use practices in situations where tenurial security means that farming is profitable to local people. Among the factors that have contributed to improved local land management are endogenous technical changes such as hill terracing that protects soil. Although this finding is close to the Boserupian scenario, the study also highlights factors that Boserup did not consider in detail, such as the inflow of ideas for technical development from exogenous sources and the reduced per capita costs of providing infrastructure as population density increases (Tiffen, Mortimore and Gichuki 1994: 267; Fairhead and Leach 1996).

The Malthusian and Boserupian views are not necessarily mutually exclusive. As Harrison notes, population growth is invariably a factor in the marginalisation of poor farmers, but the consequences of marginalisation are accentuated by inequality; the result is that '[p]opulation growth and inequality work together in a destructive synergy' (1993: 128). Where population growth is a factor in desertification it is not excessive population per se that is the problem, but skewed patterns of land ownership leading to 'hot spots' of excessive population pressure. The role of secure land tenure rights in contributing to sustainable land use is an important one, which receives some attention in Chapters 1 and 10 of this volume, as well as underlying the fencing issue contested in the case study in Chapter 6.

Inequality of land ownership is a central concern of neo-Marxist perspectives. A neo-Marxist view has some explanatory utility with respect to land degradation, although neo-Marxists have traditionally dealt with the exploitation of labour, rather than the exploitation of the environment (Redclift 1984: 122). One exception is Samir Amin who in 1977 noted that, while environmental problems accrue to all humanity, the 'centre' (meaning both the centres of the world economy and the global capitalist class) benefits most in economic terms while the periphery bears most of the costs, such as soil erosion (Amin 1977: 137–43). Other concerns of neo-Marxists are the inequities of the global economic system, such as the extraction of surpluses from periphery to centre by transnational corporations, and high international debts which put pressure on developing countries to grow cash crops for export, a factor that may place additional strain on the land as well as diverting resources from environmental conservation measures.

Piers Blaikie offers what may be termed a neo-Marxist perspective of soil erosion. Blaikie's analysis may be said to owe an intellectual debt to neo-Marxism with its emphasis on the relationship of the peasants of developing countries with the global economic system, with actors at the local, national and international levels extracting a variety of surpluses from the peasants in the form of taxes, interest and rents. These surpluses affect the cumulative land use choices of the peasantry, who need to produce more in order to subsist. As Harrison notes, the extraction of this financial surplus leads to the peasants extracting 'surplus' from the land (1993: 122). Many of the poorer peasants are confined to the poorest lands, a process Blaikie describes as spatial marginalisation (Blaikie 1985: 125). The process of spatial marginalisation is one dimension of what Mehretu, writing with respect to land degradation in Zimbabwe, refers to as spatial mismatch, namely a situation in which 'land with superior potential is comparatively sparsely populated whereas land of deficient potential is inhabited with high densities of population' (Mehretu 1995: 125). One result of spatial mismatch, or spatial marginalisation, is that the 'exhaustion of land and labour go hand-in-hand as more labour time is required to work poorer, degraded and more distant soils' (Blaikie 1985: 118). Harrison makes the same point: 'marginal people and marginal environments are chained together and become the agents of each other's destruction' (Harrison 1993: 127). Where powerful groups occupy better quality land, the options for people in more marginal areas as populations grow are restricted. Marginal populations may be forced to practice continuous cropping instead of leaving some ground fallow for a period to allow it to recover. When export demands encourage farmers to devote more of their land to cash crops rather than to maintain a diverse cropping system, the farmers become more vulnerable to bad seasons and to crashes in commodity prices. Growing cash crops for export is consistent with global market forces, though not necessarily with respect for sustainable use of the local environment.

Some African scholars have argued that environmental degradation is caused by the incorporation of African agriculture and industry into the world market. For example, to Mhone the entire southern African sub-region was a periphery of the British Empire which established a model of economic development designed principally to satisfy metropolitan interests. Mhone asserts that this legacy survives and reproduces underdevelopment throughout the region with the result 'that the human capital of the majority is deteriorating and the physical resources are being degraded and inefficiently exploited particularly in the rural sector' (1996b: 274). Dei notes the

> intensified appropriation of wealth from the rural poor majority by the state and its domestic and external accomplices (i.e. the powerful local elites and transnational corporations) and the resulting effect that the poor are unable to make the kinds of investment in land labour and capital that will help reverse the rapid deterioration of environmental conditions.
>
> (Dei 1993: 100)

These conditions, according to Dei, include desertification and soil depletion.

A neo-Marxist view explains the pattern in Africa (and the rest of the developing world) of export-led industries dominated by foreign investment and a small indigenous elite, with transnational corporations acting as the vehicles through which surplus is channelled from periphery to centre. Three case studies in this volume deal with the role of such industries in contributing to environmental degradation in Africa. In Botswana the main export-led industries are cattle and diamonds, the financial rewards from which have accrued mainly to a very small elite (Mayende 1993: 57). Both of these industries have attracted NGO campaigns against environmental degradation. With respect to the cattle industry, the best land in Botswana is occupied by the large ranchers so that poor households have been restricted to marginal drought prone arid lands. As Mhone notes, 'traditional agriculture has stagnated or even deteriorated when population growth is taken into account, and this in the face of phenomenal overall growth rates in GDP' (Mhone 1996a: 93). The little development that has occurred in Botswana has been lopsided and has reinforced dependency relations with external actors, including dependence on investment from, and migrant labour to, South Africa (Mhone 1996a: 98; Mayende 1993: 60). The NGO campaigns against the fencing of communal rangelands, which has arguably led to the over-grazing of the rangelands to breed cattle for export, and against a project to develop the Okavango Delta, which in its original form was designed in part to supply water to a De Beers diamond mine, are considered in Chapters 6 and 7 respectively. In Nigeria a long-running campaign continues against the violation by the Shell oil company and some Nigerian political elites of human and environmental rights in the Niger Delta. This campaign is analysed in Chapter 9.

Land degradation and desertification as international issues

The origins of desertification as an issue of international concern date from the severe drought and resulting loss of life in the Sahel in the 1970s. This prompted the 1977 United Nations Conference on Desertification at which a Plan of Action to Combat Desertification was adopted. The immediate objective of the Plan was 'to prevent and arrest the advance of desertification and, where possible, to reclaim desertified land for productive use'. The ultimate objective of the Plan was 'to sustain and promote, within ecological limits, the productivity of arid, semi-arid, sub-humid and other areas vulnerable to desertification in order to improve the quality of life of their inhabitants' (noted in United Nations 1991: 9). Responsibility for the Plan's coordination was assigned to the newly established United Nations Environment Programme (UNEP).

The Plan required national governments to assess desertification problems in their country, establish national priorities for action and to prepare a national plan. However, a review of progress six years later showed that anti-desertification plans had been completed by only two of more than 100 countries affected and the Special Account established by the UN to receive voluntary donations to help finance anti-desertification projects had received less than US$ 50,000. The Plan of Action was widely perceived to have had at most only limited success. If anything, the problem of land degradation seemed to become more serious after its inception. A review in 1990 identified the main shortcomings of the Plan as a lack of focus, over-ambitious goals, and lack of attention to socio-economic factors (UNEP 1991). The rural poor most immediately affected had little political leverage, rural projects were difficult to identify, plan and implement and their impact was less visible than major infrastructural projects such as dams and roads.

The proposal for an international convention to combat desertification came from the African states during the final stages of the preparatory process for the 1992 UNCED. At the time, southern Africa was experiencing a severe drought and there was resentment among the Group of 77 (G77) developing countries at the perceived northern bias of the UNCED's preparatory discussions, where attention centred on problems cast by the North as 'global issues', such as climate change, biodiversity and forests.

African countries in particular protested at the lack of attention to the issues of most salience to developing countries, such as desertification. The G77 proposed a desertification convention and at the UNCED it was agreed that a recommendation should be made to the UN General Assembly for a convention to combat desertification, with particular attention being given to Africa. The agreement to establish an International Negotiating Committee on Desertification after the UNCED may be seen as a concession to the Africans, who had otherwise received little from the UNCED process (Chasek 1994: 179). Cynics saw this agreement as a trade-off between North and South; a convention on desertification in exchange for a commitment from the G77 for a future convention on forests (Panos 1994). However, there was no straightforward trade off; the USA and eventually the European Community agreed to a desertification convention, but there was no reciprocal agreement from the South on a global forests convention (Johnson 1993: 110–11).

The perceived failure of the Plan of Action to Combat Desertification contributed significantly to the distinctive character of the Convention to Combat Desertification. In particular the criticism of the Plan of Action for not giving sufficient attention to socio-economic factors meant that they were considered legitimate topics for debate in the negotiations for the convention. The emphasis on scientific and technical 'fixes' were not as central to the debate as they were at the United Nations Conference on Desertification. The attention to socio-economic considerations was reinforced by the principles agreed at the UNCED in the Rio Declaration on Environment and Development, for example the emphasis on community participation and on eradicating poverty. There was more consideration of the fact that many of the countries suffering most from desertification and poverty also had heavy debt burdens in relation to their GNP.

Since the UNCED other international fora tasked with dealing with the environment have paid more attention to desertification, with the result that the issue receives consideration throughout the United Nations system. For example, the report of the Intergovernmental Panel on Forests to the fifth session of the Commission on Sustainable Development in 1997, noting that deforestation both affects and is affected by desertification, stated that afforestation programmes have a role in tackling desertification, soil erosion and drought. It was emphasised that forest restoration should not focus narrowly on afforestation but should also deal with the broader socio-economic context, such as the impact of trade relations on land use. The role of local and traditional knowledge, which received extensive consideration in the panel's deliberations, was also noted. However, the international dimension of global environmental degradation was reasserted, and the need for a close relationship between international instruments such as the conventions on biological diversity, climate change and desertification in addressing global environmental degradation was stressed (United Nations 1997).

The environmental debate in the late twentieth century was cast as a global crisis in which land degradation and desertification were only belatedly asserted as 'international issues'. The problems of land degradation and desertification are not new, as the Sahel crisis of 1968-73 demonstrated. However, after the 1977 United Nations Conference on Desertification the issue lost international visibility until, in the early 1990s, governments from the South sought to construct it as an international issue. However, unlike other 'international environmental issues', such as climate change, ozone depletion, marine pollution and deforestation, which involve clear transboundary ecological effects of global extent, desertification is not as yet an environmental problem with a truly global reach. The question of whether or not the problem was a global one was discussed at the negotiations for the Convention to Combat Desertification. The compromise wording in the convention is that 'desertification and drought are problems of global dimension' (United Nations 1994). The convention also notes the national dimension of the issue in calling for the preparation of national action programmes and national policies (United Nations 1994: Article 9). The local dimension is also noted, with calls for parties to 'promote awareness and facilitate the participation of local populations' (United Nations 1994: Article 5).

The emphasis in the convention on the local, national and global dimensions of desertification reflects the proprietorial nature of many land-use conflicts. For example, global forest politics are, to a large degree, characterised by competing claims that forests are a global common, a national resource and a local common (Humphreys 1996: 24-8). These claims are also discernible with respect to other forms of land use. The claims by many NGOs that natural resources should be conserved for the benefit of future generations relates, albeit often indirectly, to notions such as 'common heritage' and 'global commons'. The view that desertification is an international issue has been strengthened by the emphasis in the convention that it is a problem of 'global dimension'. This definition concealed a North–South division at the intergovernmental level between those who sought to cast the issue as a global issue (G77 governments, in particular those from Africa, who felt that internationalising the issue would strengthen claims for additional aid) and those who resisted such a definition (governments from the developed world, who wished to deflect G77 claims for additional aid). However, the internationalisation of land use is something that developing world governments are cautious of, and it is a tactic that is used only selectively. While the G77 claimed that the global socio-economic causes of desertification made the issue itself a global one, most developing world governments assert sovereignty when they perceive outside interference concerned about the international ecological effects of national land-use policies. For example, the government of Botswana has yet to agree to the Okavango Delta being classified as a World Heritage site or as a wetland of international importance under the Ramsar Convention as it refuses to accept that actors outside the country have any legitimate say on resource use in Botswana (Chapter 7).

The tension between local and national claims to land ownership is also at the heart of many resource conflicts in Africa, with competing claims being made by the state and by local communities. At the local level, indigenous peoples and local communities frequently claim that they hold customary title to local land, with many such groups, working with NGOs, asserting that they possess local knowledge essential to effective land conservation. At the national level, the state lays claim to land as a national resource. Customary common land is often sold by national governments to business concerns so that it becomes private property according to national law (Blaikie 1985: 130). Scoones notes that increasing privatisation may result in the overuse by the peasantry of remaining public lands, for example for grazing and fuelwood, as well as encroachments onto lands classified as private (Scoones 1995: 102). The result is that in many parts of Africa the status of communal lands is becoming endangered. Dharam Ghai has argued that traditional systems of resource management have been displaced when the state in Africa has taken over responsibility for managing common land previously governed by local customary rules. The result is 'uncontrolled and shortsighted exploitation of common property resources' (Ghai 1992: 11). Ghai advocates the need for 'a progressive restoration of sovereignty over natural resources to local communities' (1992: 2), but concedes that such a project faces powerful vested interests (1992: 17). Dei argues that intervention by the state on behalf of its business allies in local communities means that 'the poor find it extremely difficult to respect the customary checks, traditional beliefs and practices essential to the realisation of "sustainable development"' (Dei 1993: 102).

The different ways of framing issues relating to land degradation and sustainable resource management may have very practical implications for policy. As we will see in Chapters 3 to 9 of this book, NGOs are often involved in contesting how issues are framed as well as in lobbying for or against particular policies.

Notes

1 Haas has defined epistemic communities as 'transnational networks of knowledge based communities that are both politically empowered through their claims to exercise authoritative knowledge and motivated by shared causal and principled beliefs' (Haas 1990: 349; see also Haas 1992: 3).

2 For a comparison of the Malthusian and Boserupian views see Sage (1996: 104–5).

References

Adams, W.M. (1990) *Green Development*, London: Routledge.

Amin, S. (1977) *Imperialism and Unequal Development*, Sussex: Harvester.

Aubréville, A. (1949) *Climats, Forêts et Désertification de l'Afrique Tropicale*, Paris: Société d'Edition Géographiques, Maritimes et Coloniales.

Blaikie, P. (1985) *The Political Economy of Soil Erosion in Developing Countries*, London: Longman.

Blaikie, P and Brookfield, H. (1987) *Land Degradation and Society*, London: Meuthen.

Boserup, E. (1981) *Population and Technology*, Oxford: Basil Blackwell.

Chasek, P.S. (1994) 'From Stockholm to Rio: an analysis of 20 years of multilateral negotiation in the United Nations system', unpublished Ph.D. thesis, Johns Hopkins University, Washington DC.

Dei, G.J.S. (1993) 'Sustainable development in the African context: revisiting some theoretical and methodological issues', *Africa Development* XVIII, 2: 97–110.

Edwards, C.A. and Lofty, J.R. (1977) *The Biology of Earthworms*, London: Chapman and Hall.

Edwards, C.A. and Lofty, J.R. (1982) 'The effects of direct drilling and minimal cultivation on earthworm populations', *Journal of Applied Ecology* 19: 723–34.

Ehrlich, P. and Ehrlich, A. (1990) *The Population Explosion*, London: Hutchinson.

Fairhead, J. and Leach, M. (1996) *Misreading the African Landscape: Society and Ecology in a Forest Savanna*, Cambridge: Cambridge University Press.

Forse, B. (1989) 'The myth of the marching desert', *New Scientist*, 4 February: 32.

Ghai, D. (1992) 'Conservation, livelihood and democracy: social dynamics of environmental changes in Africa', *United Nations Research Institute for Social Development Discussion Paper* 33, Geneva: UNRISD.

Grainger, A. (1990) *The Threatening Desert: Controlling Desertification*, London: Earthscan.

Haas, P. (1990) 'Obtaining environmental protection through epistemic consensus', *Millennium: Journal of International Studies* 19, 3: 347–63.

Haas, P. (1992) 'Introduction: epistemic communities and international policy coordination', *International Organization* 46, 1: 1–35.

Hare, F.K. (1977) 'Climate and desertification', in United Nations, *Desertification: Its Causes and Consequences*, Oxford: Pergamon.

Hare, F.K. (1985) *Climatic Variations, Drought and Desertification*, Geneva: World Meteorological Organization.

Harrison, P. (1993) *The Third Revolution: Population, Environment and a Sustainable World*, London: Penguin.

Hellden U. (1991) 'Desertification: time for an assessment', *Ambio* 20, 8: 372–83.

Hulme, M. and Kelly. M. (1993) 'Exploring the links between desertification and climate change', *Environment* 35, 6: 4–11 and 39–46.

Humphreys, D. (1996) *Forest Politics: The Evolution of International Cooperation*, London: Earthscan.

Johnson, S.P. (ed.) (1993) *The Earth Summit: The United Nations Conference on Environment and Development (UNCED)*, London: Graham and Trotman.

Jones, B. (1938) 'Desiccation and the West African colonies', *Geographical Journal* 41: 401–23.

Leach G. and Mearns, R. (1988) *Beyond the Woodfuel Crisis: People, Land and Trees in Africa*, London: Earthscan.

Mainguet, M. (1994) *Desertification: Natural Background and Human Mismanagement*, Berlin: Springer-Verlag.

Mayende, G.P. (1993) 'Bureaucrats, peasants and rural development policy in Botswana', *Africa Development* XVIII, 4: 57–78.

Mehretu, A. (1995) 'Spatial mismatch between population density and land potential: the case of Zimbabwe', *Africa Development* XX, 1: 125–46.

Mhone, G. (1996a) 'Botswana economy still an enclave', *Africa Development* XXI, 2 & 3: 89–99.

Mhone, G. (1996b) 'The socio-economic crisis in Southern Africa (Botswana, Malawi, Zambia, Zimbabwe)', *Africa Development* XXI, 2 & 3: 267–78.

Morgan, R.P.C. (1986) *Soil Erosion and Conservation*, Harlow: Longman.

Mortimore, M. (1993) 'The Sahel', in G. Sjostedt (ed.) *International Environmental Negotiation*, California: Sage.

Morris, J. (1995) *The Political Economy of Land Degradation: Pressure Groups, Foreign Aid and the Myth of Man-Made Deserts*, London: Institute of Economic Affairs.

Munslow, B., with Katerere, Y., Ferf, A. and O'Keefe, P. (1988) *The Fuelwood Trap: A Study of the SADCC Region*, London: Earthscan.

Panos (1994) *Panos Media Briefing* (8), June, London: Panos.

Redclift, M. (1984) *Development and the Environment Crisis: Red or Green Alternatives?*, London: Routledge.

Sage, C. (1996) 'Population, poverty and land in the South', in P. Sloep and A. Blowers (eds) *Environmental Policy in an International Context: Environmental Problems as Conflicts of Interest*, London: Arnold.

Scoones, I. (ed.) (1995) *Living with Uncertainty: New Directions in Pastoral Development in Africa*, London: Intermediate Technology Publications.

Scoones, I. and Toulmin, C. (1999) *Policies for Soil Fertility Management in Africa: A Report Prepared for the Department for International Development*, London/Brighton: International Institute for Environment and Development/Institute for Development Studies.

Stebbings, E.P. (1935) 'The encroaching Sahara: the threat to the West African Colonies', *Geographical Journal* 85: 506–24.

Thomas, D.S.G. and Middleton, N.J. (1994) *Desertification: Exploding the Myth*, Chichester: Wiley.

Tiffen, M., Mortimore. M., and Gichuki, F. (1994) *More People, Less Erosion: Environmental Recovery in Kenya*, Chichester: Wiley.

Toulmin, C. (1993) *Combating Desertification: Setting the Agenda for a Global Convention*, London: International Institute for Environment and Development.

UNEP (1991) *Status of Desertification and Implementation of the United Nations Plan of Action to Combat Desertification*, Nairobi: United Nations Environment Programme.

UNEP (1992) *World Atlas of Desertification*, London: Edward Arnold.

United Nations (1977) *Desertification: Its Causes and Consequences* (compiled and edited by the Secretariat of the United Nations Conference on Desertification), Oxford: Pergamon.

United Nations (1991) 'Protection and management of land resources', UN document A/CONF.151/PC/62, 22 July.

United Nations (1993) *Agenda 21*, New York: United Nations.

United Nations (1994) *Elaboration of an International Convention to Combat Desertification in Countries Experiencing Serious Drought and/or Desertification, Particularly in Africa: Final Text of the Convention*, A/AC.241/27, New York: United Nations.

United Nations (1997) 'Note by the co-chairmen of the Ad Hoc Intergovernmental Panel on Forests', UN document E/CN.17/IPF/1997/3, 3 January.

3 NGO influence at the International Negotiating Committee for a Convention to Combat Desertification

Perspectives from Zimbabwe and Nigeria

Susan Carr, Alan Thomas, Enoch Okpara, Roger Mpande and David Humphreys

The Convention to Combat Desertification (CCD) was the first international environmental convention to be agreed since the 1992 UNCED in Rio de Janeiro. Five substantive negotiating sessions of the Intergovernmental Negotiating Committee on Desertification (INCD) were held in 1993–4. A further five sessions were held in the 'interim period' between the conclusion of the negotiations of the convention in June 1994 and its entry into legal effect in December 1996.[1] This chapter analyses the negotiating sessions of the INCD as a target for NGO influence. Two of the authors represented sub-Saharan African NGOs at the INCD, namely Mpande of the Zimbabwe Environment Research Organisation and Okpara, of the Nigerian Environmental Study/Action Team (NEST). Carr and Thomas also attended with observer status. Evidence was collected at the national level in Nigeria, Zimbabwe and the SADC countries and at INCD negotiating sessions.

The investigations were informed throughout by three questions. First, what provisions in the CCD were included specifically as a result of NGO influence? Second, to what extent, and in what ways, did the process differ from that in other UN conferences or other international arenas as a result of the influence of NGOs? Third, what new influence or role do NGOs have in putting the CCD into practice, compared to other international agreements?

The negotiations and the role played by NGOs

Various contextual factors contributed to the distinctive character of the INCD agenda. The emphasis in Agenda 21 on local communities and indigenous peoples was a factor, as was the fact that the convention was proposed by the developing rather than the industrialised countries. Criticism that the Plan of Action to Combat Desertification (Chapter 2 of this volume) did not give sufficient attention to socio-economic factors meant that they were considered legitimate topics for debate in the negotiations for the convention. Hence the guidelines set out in UN General Assembly Resolution 47/188 (December 1992) stressed the need to involve local people fully in decisions and to respect indigenous knowledge and ensured that the negotiations were not dominated by scientific and technical matters. Attention to socio-economic considerations was reinforced by the principles agreed in the Rio Declaration on Environment and Development, in particular the emphasis on poverty eradication. As INCD Chair Bo Kjellén of Sweden commented before the negotiations, the desertification issue 'is very much an example of the clear link between development and environment and it is an issue where the very notion of poverty is at the centre' (CCF 1993: 14). Finally, the negotiations were seen by the Group of 77 developing countries (G77) as a further opportunity to promote their UNCED concerns on external debt and North–South inequities. Here it is noteworthy that the INCD Secretariat answered directly to the General Assembly. Theoretically this gave greater power to the needs of the G77 which is in a majority in the General Assembly (Porter and Brown 1991: 50–1).

Economic national interests play a key role in multilateral conference diplomacy as institutionalised in the United Nations and its specialised agencies. A problem-solving negotiating process is generally considered to be the best procedure to reach agreement in such negotiations rather than a diplomatic bargaining process under zero-sum conditions (Holsti 1977: 186, 209). Problem-solving involves two or more parties attempting to develop rules to handle a problem arising in the physical or diplomatic environment. It involves a common study of a problem, pooling of resources and cooperation for mutual benefit with no party sacrificing major values in the process.

Characterised by technical research, analysis and data, the procedures adopted for the INCD are consistent with problem-solving negotiations. Delegations required knowledge of the various dimensions of the subject matter, hence many delegates were drawn from the Ministries of the Environment or relevant UN agencies. The INCD adopted procedures to enhance the exchange of scientific, technical and socio-economic information among delegations at plenary sessions. These included the following.

First a multidisciplinary International Panel of Experts on Desertification was established to provide data and information on the socio-economic causes and consequences of desertification in Africa. The Panel had fifteen members who acted in an individual rather than representative capacity. One such individual was from the UK-based International Institute for Environment and Development (IIED). The Panel of Experts met before each negotiating session. Second, an inter-agency group was created to coordinate the scientific and technical knowledge of relevant UN Agencies including the United Nations Environment Programme, United Nations Development Programme, World Meteorological Organization, United

Nations Sudano-Sahelian Office and the Food and Agriculture Organization. Third, case studies were commissioned with terms of reference on dryland management and anti-desertification programmes using funds made available by developed country members.

As with other international conventions, the negotiation of the CCD can be divided into three phases: issue definition; exchange of statements on initial positions; and negotiating and forging consensus on the text. However, there was one fundamental difference between the INCD negotiations and other multilateral negotiations where the first phase, which defines the scope and magnitude of the problem, takes place before the main negotiations, for example at a pre-negotiation conference or within a UN agency (IISD 1993; Chasek 1994). In the case of the INCD, issue definition was not considered until the first INCD session.

NGOs played an important role in this phase. UN General Assembly Resolution 47/188 invited all relevant NGOs, especially those from developing countries, to attend the negotiations. NGOs were accredited to the INCD upon presentation of their credentials. By the first session 173 NGOs were accredited to the INCD, with more than 100 representatives from forty-eight NGOs attending. By the fifth session the number of accredited NGOs had almost doubled, to 300, with over fifty NGOs actively taking part. Many were from Africa, including local grassroots NGOs whose representatives had never attended an international negotiating session. The procedural rules for NGO involvement in the negotiations were similar to those for the UNCED; NGO representatives were allowed to deliver statements with the permission of the chair.

Kenya Energy and Environmental Organizations (KENGO) took the initial lead in coordinating NGO activity at the first INCD session. KENGO held a meeting of Kenyan NGOs to contribute to the preparations of the Kenyan delegation and a three-day workshop for African NGOs. The Environment Liaison Centre International (ELCI), based in Nairobi and therefore close to UNEP, organised a two-day meeting with African and non-African NGOs. This resulted in a joint NGO statement to the INCD on the relationship between biodiversity, climate change, the inequitable distribution of resources, and the impact of international trade, external debt and colonialism on land use. Eventually ELCI prevailed over KENGO as the NGO focal point in the RIOD (Réseau International d'ONG sur la Désertification) NGO network on desertification established at Ouagadougou (November 1994). RIOD served as an organisational focus for the NGOs, although it had no formal leadership position. Indeed there was no defined leadership structure among the NGOs, although an informal system of participatory leadership and coordination was adopted. Members volunteered for tasks such as chairing NGO Committees, monitoring debates in working groups and making advance preparations for INCD sessions.

The first phase of the negotiations, identifying and defining the problem, and the second, exchanging statements of initial positions, took up most of the first session and part of the second. Some delegations emphasised the degree of scientific uncertainty on the natural and socio-economic causes and consequences of desertification, leading to conflict between delegations on issue definition. Consensus was reached on the idea of national and sub-regional action programmes, the need for improved research and development and the transfer and adaptation of technology. The main source of disagreement at this stage was whether drafting a regional annex for Africa should take priority over the annexes for other regions. There was also controversy concerning international financial commitments, which African and some other developing countries argued should come from the developed countries. The developed countries argued that these commitments should come principally from the countries affected.

Despite a degree of NGO unity represented by a joint statement, the first INCD session resulted in divisions between African and non-African NGOs. Some representatives of African NGOs wanted non-African NGOs excluded from their meetings. When the negotiations stalled over the priority to be given to the drafting of the four regional annexes, NGOs tended to divide along regional lines, often supporting the position of their own countries. This rift on procedural matters limited the political space for lobbying on more substantive issues. To provide an opportunity for NGOs to share their experiences and agree concrete proposals before the second session of the negotiations, the INCD Secretariat, with funds from the Dutch government and other donors, set up an NGO meeting at Bamako in Mali, which was attended by over 100 representatives from all over the world. Participants presented case studies of NGO involvement in anti-desertification measures, although little attention was given to developing a coordinated NGO lobbying strategy. In this respect NGOs attended the second session as ill-prepared as at the first.

At the second session the debate on the definition of desertification was eventually brought to a close. Contention had centred on whether or not desertification was a global issue. The G77 warned against demoting the convention to a second-class international legal instrument, arguing that it should have the same status as the climate change and biodiversity conventions. To strengthen their case that desertification was a global problem, the G77 introduced arguments made in the climate change negotiations and, in so doing, sought to shift some of the responsibility for desertification to industrialised countries. The intention was to place developing countries in a stronger position to demand assistance from other countries (Toulmin 1993). The developing countries argued that desertification is caused in part by anthropogenic climate change. The rationale for this strategy was that the industrialised countries had accepted that historically they had 'common but differentiated responsibilities' for climate change due to their higher rate of greenhouse gas emissions (United Nations

1992a: Article 3.1). The principle of common but differentiated responsibilities also appears in the Rio Declaration on Environment and Development which further notes the responsibility of the developed countries 'in view of the pressures their societies place on the global environment and of the technologies and financial resources they command' (United Nations 1992b: Principle 7).

The G77 also sought to establish a relationship between desertification and global economic phenomena such as external debt and the perceived unfavourable structure of international trade. The G77 had introduced these issues into the UNCED forest negotiations where the developed countries were reluctant to recognise them as salient to the negotiations (Humphreys 1996: 83–103). The developed countries were similarly disinclined to recognise these issues in the INCD negotiations. Denying that there was a global dimension to desertification, the USA and Australia pointed out that they had lived with desertification for decades and that international action was not necessary to deal with what they considered to be localised, land management problems. The developed countries cautioned that the developing countries should not expect much with respect to new and additional financial resources. The EC, supported by Canada, wanted the convention to focus on specific desertification problems and to avoid general references such as North–South inequities and the eradication of poverty. Eventually a compromise definition was agreed, so that the final text reads: 'Acknowledging that desertification and drought are problems of global dimension in that they affect all regions of the world …' (United Nations 1994).

During the second session the NGOs began to make their presence felt among the delegates, although the lack of substantive negotiation remained a source of frustration for some. Few northern NGOs attended and it became clear they did not consider the convention a priority. After the second session the International Development and Research Centre (IDRC) in Canada organised regional workshops on themes relevant to the negotiations, such as traditional knowledge and structural adjustment programmes. These workshops allowed NGOs to share their experiences and position themselves on each of the issues.

The third phase of the negotiations – negotiating and reaching consensus on the main body of the text – continued throughout the third, fourth and fifth sessions. By this time the NGOs were better organised and more involved in the negotiations. They agreed a programme of action for lobbying delegates and each NGO was assigned specific tasks. Meetings were held with delegations and intergovernmental organisations. NGOs also attended the regional meetings of the African delegates and the G77. Meetings with the Chairman and Executive Secretary helped clarify procedures for NGO participation and gave NGOs the confidence to contribute to the negotiations. On several occasions statements made by NGOs were fully endorsed by delegations. To maintain the momentum

NGOs agreed to establish four working groups to address different issues between sessions.

Unresolved issues at the end of the third session included the scope of the convention, namely whether it should cover international economic issues such as external debt, trade and poverty, and obligations and commitments, in particular whether rapidly developing countries such as Malaysia should be included among the donors. G77 proposals for new international institutions to implement the convention were resisted by developed countries.

At the fourth session, NGOs felt they were able to make a real contribution to the negotiations. The NGOs began to look ahead to the implementation of the convention and to plan the creation of an NGO desertification network. Their suggestion that national desertification trust funds managed by representatives from government, donors, researchers and the community should be established to channel donor funds was endorsed by delegates and a task force was formed to develop further this idea.

At the fifth session, NGOs lobbied delegates to guard against a weakening of the convention in the final stages. By this time a cooperative style of working had emerged among the NGOs. Despite problems, such as accommodating new NGOs at each session, continued funding uncertainties and the impossibility of forging NGO consensus on all issues, NGO unity was built on clear common goals and well-defined lobbying and communication tasks. However, there was less unanimity on proposals for the NGO desertification network.

At the end of the fifth session, after three all-night negotiating sessions to resolve remaining difficulties, delegates adopted an agreed text (including regional annexes for Africa, Latin America and the Caribbean, Asia, and the Northern Mediterranean) within the original schedule. As is usual for initial agreements in a new area (Sjöstedt 1993), the agreement is quite weak in several respects. The final text includes only weak commitments from donor countries on extra funding, and emphasises instead the need for better coordination and more effective use of existing funds. Developing countries made commitments to adopt more participative approaches, but there is no indication of how this is to be achieved. No clear targets or indicators of progress were agreed. Contentious issues such as terms of trade, and the relationship between global warming and drought, although mentioned in the preamble, were weakened with caveats by the developed countries who argued that they should be dealt with by other international negotiations.

The contribution of African NGOs to the convention

African NGOs played a significant national awareness-raising role on desertification from the start of the INCD process. In a survey for ELCI of official and NGO delegates to the seventh INCD session from thirty-six countries, Scherl (1996) found that by August 1995 awareness-raising activities involving NGOs had begun in twenty-one countries (58 per cent of the sample), NGOs were involved in the development of National Action Plans in ten cases (28 per cent), and NGO national committees had been established in fifteen countries (42 per cent). The cases of Zimbabwe and Nigeria are used to illustrate NGO activities in two African countries.

Zimbabwe is a member of the Southern African Development Community (SADC).[2] Despite the urgency of serious drought in the SADC sub-region in 1991–2, no more than two NGO representatives from the sub-region attended an INCD meeting up to INCD 7, with ZERO (Zimbabwe Environmental Research Organisation),[3] a regional environmental research organisation based in Harare, usually present. However, with no effective feedback mechanism for the few participating NGOs, there was a lack of awareness about the convention amongst most NGOs and CBOs (community-based organisations) in the sub-region.

The Interim CCD Secretariat commissioned Roger Mpande, a member of ZERO and one of this chapter's co-authors, to prepare a discussion paper on NGO and CBO involvement in the SADC Subregional Action Programme on Drought and Desertification for a sub-regional NGO meeting in Windhoek, Namibia in January 1996. The paper used the guidelines of RIOD's NGO Action Plan to propose: a mechanism for dialogue among SADC NGOs and CBOs working on drought and desertification; ways of raising financial and other resources for implementation; identification of transboundary projects requiring NGO or CBO participation; strengthening relevant existing NGO and CBO networks; lobbying by NGOs and CBOs on relevant socio-economic factors such as land reform and resettlement; cooperation among stakeholders on monitoring the Action Programme; and the involvement of universities and other institutions.

The Windhoek meeting agreed recommendations on the urgency of creating 'an enabling environment at higher levels to facilitate actions at local and national levels'. The CCD stresses the national and local levels for implementation, as do the NGOs and CBOs involved. The Windhoek meeting also raised awareness of the CCD at the sub-regional level judging by the increased number of SADC NGOs at INCD 8 where, in addition to Zimbabwe, Namibia, Botswana, Tanzania and South Africa were represented. However NGOs so far have a minor role in most SADC countries, and if this is to change it is likely to be through the efforts of governments and international agencies promoting NGOs rather than from NGO agitation.

At the national level in Zimbabwe there has been considerable activity in government and amongst NGOs. The government established a consultative National Forum in November 1995 to enable stakeholders to define the agenda, expectations and responsibilities for the development of the National Plan of Action. The stated intention was that the meeting should serve as a first stage in the grassroots evolution of the plan, although government and donor agencies were strongly represented. Departmental responsibility for the plan rested with the Ministry of Environment and Tourism. The National Advisory Committee on Sustainable Development (NACSD) acted as a steering committee while the National Action Plan was prepared. The government also commissioned studies by NGO experts and foreign consultants to advise on coordination mechanisms for community participation.

The convention can be seen in terms of opportunities balanced by obligations. Despite the lack of commitment from donor countries to provide new financial resources, there is a window of opportunity in which countries like Zimbabwe can expect assistance from the international community for projects developed under the convention. In return, the convention legally obliges recipient governments to integrate the dryland populations and NGOs as active partners in the planning and implementation of natural resource management in arid and semi-arid areas. There are obligations to establish long-term integrated strategies using a 'bottom-up' participative approach to combat desertification and for contingency plans in affected areas. Guidelines are available on how to carry out the required coordination and consultation (e.g. UNDP 1995). Setting up the National Forum and NACSD are the first steps in this process.

The situation in Nigeria is very different from that of Zimbabwe owing to political instability and frequent changes of government. In its draft 1996 Country Strategy Note, the government acknowledges this constraint by noting that 'governance and policy instability... have not provided a conducive environment for economic and social activity to progress at the desired level' (FRN 1996: 15).

In 1993 the Nigerian Federal Environmental Protection Agency (FEPA) set up an inter-ministerial National Committee on the Desertification Convention. Some NGOs, including the Nigerian Environmental Study/Action Team (NEST), were invited to make an input into the national negotiating document. Enoch Okpara, one of this chapter's co-authors, attended the INCD representing NEST. A leading Nigerian scientist was appointed on the International Panel of Experts on the Desertification Convention. After the first INCD session in Nairobi, where Nigeria's presence was noticeable, the country made only limited appearances in subsequent negotiations. Although Nigeria signed the convention in October 1994 no preparations were made, as required by the convention, to draw up a National Action Programme (NAP) during the 'interim period'.

Many of the environmental policies of the Nigerian government formulated prior to the INCD process were 'top-down' and were criticised by Nigerian NGOs for shortcomings arising from inadequate consultations among affected actors. One such example is FEPA's 'National Policy on the Environment', a package of policies on development and the environment.

On land use and soil conservation, for example, the policy provides for fifteen strategies to be adopted including 'developing through research, sustainable agro-forestry techniques, for checking the spread of desertification' (FEPA 1989: 7). Although Nigeria's 1996 Country Strategy Note mentions a number of international conferences, including the Beijing Declaration, the International Conference on Population and Development and Habitat II, it is silent on the desertification convention (FRN 1996).

Four explanations may be offered for the apparent disinterest shown in the desertification convention by the Nigerian government at the time of the negotiations. First, and as already noted, there were preoccupations with state security and the transition to democracy. Second, the fact that the Sahelian region was not currently witnessing any devastating incidence of drought meant that the INCD negotiations were not high on the Nigerian national environmental agenda. Third, Nigeria has initiated anti-desertification programmes and projects outside the desertification convention framework, including sand dune stabilisation, shelter-belts and community woodlots, afforestation programmes, roadside tree planting, raising and distribution of seedlings and community development projects in desertification-prone environments. Some of the projects are co-funded with the EU. The Forestry Management, Evaluation and Coordinating Unit (FORMECU), an arm of the Nigerian Federal Department of Forestry, has, in collaboration with the World Bank, adopted afforestation as a mechanism for minimising desertification. The programme emphasises research in vegetation and land-use management, arid land human ecology and rangeland and soil management. Fourth, desertification in Nigeria is relatively localised in the national context. There is a tendency in Nigeria to address environmental problems at the state level as opposed to the federal (national) level, except where they attain the status of a national disaster. Hence it has proved difficult to mobilise federal resources to address a problem that affects only the northern states of the federation.

Given this policy environment, effective NGO advocacy on the desertification convention was difficult. The major constraints facing Nigerian NGO activity in the INCD are threefold. First, many of the NGOs that were part of the National Committee on the Desertification Convention could not follow up the process at the international level due to lack of sponsorship. Second, most of the NGOs perceived that their activities were not directly related to desertification and that they did not have a stake in the INCD. This perception was reinforced by the absence of a national debate on desertification to which NGOs could contribute. Third, there are fewer NGOs (as distinct from CBOs) in the northern Sahelian zone of the country affected by desertification compared to the south. The tendency is for the southern-based NGOs (with NEST and the Nigerian Conservation Foundation exceptions) to address environmental problems peculiar to their own areas. Desertification and drought are not seen as a part of their mandate. In the absence of logistic and operational support, it is virtually impossible for any NGO to address meaningfully an environmental problem afflicting an area several hundreds of kilometres from its operational base, except through intermediary organisations.

Despite these constraints, NEST was able to sensitise ten other NGOs who had indicated an interest in working on desertification or related issues, through letters and an advocacy publication. The advocacy publication was widely circulated to nearly 1,000 people, including community change agents, policy makers, the media, development practitioners and NGOs. Visits were also made to the Federal Environmental Protection Agency (FEPA), in Abuja, with a view to encouraging government to elaborate a policy on combating desertification.

African NGOs and the implementation of the convention

A big role for NGOs has been written into the convention on implementation. NGOs are to help facilitate the community-level implementation central to the CCD, assist with the proper use of indigenous knowledge and play a role in the new national and international institutions. National Action Plans are to be drawn up and continuously updated using consultative processes. Cooperation mechanisms at the local level will include NGOs. However, it is unclear how far African NGOs will be able to live up to the hopes invested in them. Action by some African NGOs may be hindered by opposition from governments. In international law a bottom-up approach may challenge certain notions of sovereignty, while some of the more specific innovative elements in the CCD, such as the idea for national trust funds with NGO involvement, may conflict in detail with the perceived interests of some governments, even those of states which support the convention.

Nevertheless, the importance of implementation cannot be underestimated as a source of NGO influence. Before the recent wave of interest in NGOs, Grindle (1980) noted how the implementation process is especially central to developing countries. Charlton and May have argued that NGO involvement in implementation has an important cumulative political impact in developing countries (1995: 241). Thus, although NGO influence in the CCD has mostly been restricted to setting the agenda and providing information and ideas, the expectation that they will also assist with implementation could lead to considerable future influence. Certainly a lot of government activity is aimed at preparing new institutional arrangements. Furthermore, when the convention was adopted, the INCD passed a Resolution on Urgent Action in Africa, under which governments of affected states and their donor partners were encouraged to start joint activities.

Following the negotiation of the convention, a further five INCD sessions were held (see note 1). The sixth INCD session agreed a work programme to lead up to the first meeting of the Conference of Parties and defined mandates for the INCD's two working groups. At the seventh session these working groups began their work. Working Group I had four issues to consider: the identification of an organisation to house the 'global mechanism'; the designation of a permanent secretariat; draft financial rules for the Conference of Parties; and the programme and budget. Working Group II covered the organisation of scientific and technical cooperation, draft rules of procedure, and procedures for communicating information and reviewing implementation.

Perhaps realising the enormity of the challenges ahead, during the seventh and eighth sessions many delegates expressed dissatisfaction at the organisation, frequency and siting of INCD sessions, at the large amount of time allocated to Africa and the lack of priority given to desertification outside the INCD by donor countries and even some affected countries. There was discussion on scientific and technological cooperation, including how to keep the expert Committee on Science and Technology (CST), and its bureau and ad hoc panels, small and efficient, while ensuring geographic, multidisciplinary, gender, NGO and international agency representation. The difficulty of making progress once negotiations moved from agreeing on general principles to implementation and resources was apparent. Little progress was made on the global mechanism and financial rules. The NGOs were notably less harmonious during the seventh and eighth sessions compared to the joint working style achieved from the third to the fifth sessions. The NGOs also spent considerable time debating the location of the focal points for the RIOD network, resource allocations for attendance at international meetings and the coordination and regulation of funding applications by RIOD members. Nevertheless, NGOs made significant inputs and pressed for the CST to include nominees from NGOs and for the outputs of the CST's deliberations to be communicated to the peoples of the drylands.

Perhaps the most significant progress was made outside the official negotiations. Several of the multilateral agencies promoted the convention by holding working parties or conferences and developing policies in the context of the CCD framework provided. The main UN agencies working on desertification are UNEP, UNDP and the UN Office to Combat Desertification and Drought (formerly UNSO, the UN Sudano-Sahelian Office). Other agencies within and outside the UN, such as FAO and IFAD, are also heavily involved. The World Bank announced in 1995 'an additional $2.0 billion in investments directed towards desertification problems over the next three years' (World Bank 1995). Many of these agencies have consulted NGOs as part of their policy development, such as UNSO in its work on national desertification funds, an idea largely promoted by NGOs and now adopted at UN agency level with assistance from ELCI and other NGOs. Commitments to consultation and participation are included in UNDP's report on building partnerships to support implementation at national level (UNDP 1995) and FAO's conceptual framework for effective community participation (FAO 1993).

At the NGO meeting in Burkina Faso (November 1994) when RIOD was formally established, an NGO Action Plan to Combat Desertification was adopted. This includes guidelines for organising an NGO Forum and a national NGO coordinating committee on desertification (NCCD) and for NGOs and CBOs to take the initiative in forming mechanisms for community consultations for National Action Plans. The meeting agreed a network of 'focal points' through which RIOD would operate from global to regional and sub-regional levels, with different NGOs having responsibility for coordinating communication at each level of the network. Guidelines were agreed for the procedures through which each NCCD should select an NGO to act as the national focal point in the RIOD network.

The problem remains of how to implement a 'bottom-up' participatory approach at provincial, district or local levels. Zimbabwe, for example, is relatively well placed in this respect since it has a plurality of institutions at each of these levels. Some coordination mechanisms are already in place (though not always working effectively). There are several projects in Zimbabwe to arrest or reduce desertification which, while the responsibility of particular ministries, are in practice undertaken by the communities themselves, generally without knowledge of the CCD. Zimbabwean national programmes on land reform and decentralisation have tended to reach communities via a 'top-down' approach through national, provincial and district levels and down to wards and villages. Community participation has often meant in practice that local people are asked to participate in a series of activities determined by government or external agencies.

The question of whether power is located at the national or local levels is therefore central. But as Toulmin argues, 'a sensitive issue such as power cannot be tackled directly by a consensus document such as a Convention, but its neglect risks the promotion of policies which assume a degree of harmony which is unrealistic' (1997: 61). Decentralisation may simply replace conflicts of interest at the national level with those at the local level. Almost all governments and agencies now espouse a participatory approach, but often find it difficult to accommodate such an approach within existing frameworks. Indeed some have adopted the language of participation as the price for more funding, and the fact that some governments at least have adopted the CCD without thinking through its implementation can lead 'to a lack of respect for international law because such treaties are signed and ratified without a clear understanding of the problem' (Toulmin 1997: 62–3).

One proposal from NGOs in Zimbabwe was to use the existing District Environmental Action Plan process (DEAP). Initiated in 1995 by the Department of Natural Resources (DNR) within the Ministry of Environment and Tourism, DEAPs operate at Rural District Council (RDC) level, are supported by UNDP with technical assistance from the World Conservation Union (IUCN), include concern for environmental protection, and emphasise resource utilisation through participatory rural appraisal. The proposal was to develop the National Action Plan to Combat Desertification within the framework of the DEAP process and under the co-ordination of the DNR, the status of which within government would have to be upgraded.

In Zimbabwe's case there is a good institutional framework for implementing environmental programmes, but further strengthening of co-ordination mechanisms is necessary if NGOs are to play a role in the implementation of community programmes and in lobbying government to ensure it maintains an enabling policy environment for the realisation of the National Action Plan. In other countries the role of NGOs in implementation may be correspondingly even more difficult. In Nigeria, for example, there are very poorly-developed links between federal government, states and the grassroots, and local government has very low organisational capacity. However, in Nigeria as in Zimbabwe, there is now increased recognition of the urgency of the problems caused by land degradation and the need for policy ideas. Hence there may be windows of opportunity for NGOs if they can offer solutions, either within or outside the CCD framework.

It is hard to overestimate the scale of organisational change required in many countries before national-level coordination and effective participation of local communities and farmers into decision making can come into effect. Some of the obstacles include the need to communicate with a wide range of actors, to train staff in new skills and to break down traditional departmental or disciplinary barriers. Even basic practical matters like the availability of transport to take staff to rural areas may present problems. Where NGOs offer or are invited to act as intermediaries between government officials and local communities, they may find their resources and capacity overstretched. The funding arrangements of many donor agencies are geared to handling large-scale projects, often with visible returns to the donor country. The donors may have considerable difficulty identifying reliable and effective NGOs to support, and may prefer to channel funds for grassroots activities via one or two 'apex' or umbrella NGOs, preferably with strong international connections. For all these reasons, NGOs will need to continue to press for increased participation and to monitor all aspects of the CCD.

An analysis of NGO influence on the negotiation of the convention

Eight points can be made about the overall influence of NGOs on the negotiation process for and provisions of the Convention to Combat Desertification. First, as noted earlier, the resolution establishing the INCD invited NGOs to join the negotiating process. NGOs have always had an impact on the UN, and provision is made for them in the UN Charter. However, as a result of the UNCED there have been changes both in the way NGOs are accredited and in the roles they can play. Before the UNCED, most NGOs accredited to the UN were international NGOs, whereas accreditation has now become more open to local, grassroots and national NGOs. The precedent of the UNCED, where NGOs were given access upon presentation of their credentials, was used at the INCD sessions. As Willetts notes: 'Now it is possible to go directly from the grass-roots to the UN, by-passing both governments and the complex structures of international NGOs' (1996: 79). Finger (1994) suggests that the UNCED model for NGO involvement is one in which they should have a lobbying rather than simply a consultative role. In other words, NGOs should expect to influence the negotiations rather than content themselves with a chance to input information and ideas. Due to the lack of experience of many NGOs, this was not fully appreciated by the NGO community at the INCD until relatively late in the negotiations.

Second, the whole ethos of the CCD and the way the desertification issue was defined provided NGOs with more opportunities to influence the outcome than at previous international negotiations. Because the lead was taken by developing countries, there was far greater emphasis on cross-cutting issues such as trade and poverty compared to environmental conventions negotiated before UNCED. From the outset it was recognised that the full involvement of local people would be essential. The practical experience of NGOs in working with local people was acknowledged.

Third, the outcome of the negotiations was relatively successful for NGOs. The final clauses of the CCD correspond closely to NGO priorities, including participation, community-level implementation and indigenous knowledge. The proposed new institutions, such as National Action Plans and the 'global mechanism' to help mobilise funds and international research all envisage important roles for NGOs. According to Lean:

> The Convention breaks new ground by enshrining a bottom-up approach in international law. It repeatedly emphasises the importance of full participation, and specifically underlines 'the important role played by women'. It also stresses 'the special role of non-governmental organisations' and gives them an important role in ensuring implementation.
>
> (Lean 1995: 15)

A fourth, more speculative, point relates to the style of the negotiations. The presence of NGOs appeared to lead to more informality and more openness in information gathering and sharing. African NGOs were especially prominent, alongside a small number of mostly specialist northern NGOs. This was partly the result of the absence of many large northern NGOs. Many of the African and other NGOs present are used to an informal, participative style of working, and to seeking innovative solutions to problems in collaboration with others, and certainly something of this style was to be found in the workings of the INCD. Although delegates often reverted to a more adversarial form of official negotiation, the shared understanding that they needed to reach agreement led at other times to a joint problem-solving style that was closer to that of the NGOs.

Fifth, despite the expectations on them to be more active, the NGOs in the INCD failed to take full advantage of the opportunities offered to them. Their role was mostly restricted to a consultative one. Northern NGOs with experienced lobbyists and substantial research and organisational capacity, such as the World Resources Institute (WRI), the World Conservation Union (IUCN) and the World Wide Fund for Nature, played little or no active role in the negotiations. Most of the NGOs present lacked lobbying experience and were not well coordinated in the critical early INCD sessions, hence their influence was restricted to certain phases of the negotiation process.

Haufler (1993) suggests that the relations between states and non-state actors tend to take one of two forms. In the first, the states are dominant, using NGOs to provide ideas about practices and norms and then for help in implementation. In the second, either the roles are reversed or states and non-state actors play a relatively equal role. Governments and intergovernmental organisations prefer NGOs to play the first role (Finger 1994: 186), and this was clearly the case with NGOs in the INCD. Hence, although they provided information and ideas, NGOs were not prominent or influential when there were important policy choices to be made. The main areas of contestation, noted above, were disagreements between states or groups of states with differing national interests. It is possible to explain why agreement was reached using interest-based regime theory with little or no reference to NGOs (Carr and Mpande 1996). According to this view, the open-door policy for NGOs at the INCD served the purposes of states, since it provided fresh perspectives, but reduced the likelihood of coordinated NGO lobbying on any substantive issue other than the need for greater NGO involvement.

Thus the fifth point is that on the evidence one cannot conclude that NGO influence during the negotiations was the driving force behind the new elements in the CCD, even those which formed part of the NGOs' agenda. On the contrary, it appears that the NGO role was enhanced by the organisation of the negotiating sessions and by the INCD Secretariat rather than by NGOs themselves. Our research reveals some disappointment in the Secretariat that the NGOs did not achieve more. Although the broad definition of the issue made it possible for NGOs to raise many of their concerns, the range of NGOs involved made it difficult to establish a coherent NGO position. However, topics raised by NGOs that had some support among government delegates were not easily dismissed. For instance, when KENGO, speaking on behalf of all the NGOs, raised the issue of land tenure reform, the Kenyan delegation responded that the convention should not interfere with national government policies. But subsequently the convention's executive secretary stated that the use of sovereignty arguments to prevent debate about issues fundamental to the convention demonstrated a country's lack of commitment to the negotiating process and would be recorded as such. In many areas NGOs were pushing against an open door.

This leads to the seventh point, that in many respects the more important influence was not that of the NGOs on the negotiations, but the reverse; the influence of the negotiations on the NGOs. NGOs played a major role in part because of expectations on them. Especially for grassroots African NGOs, the negotiating process provided a valuable opportunity to establish networks and to build up their capacity for both coordinated and individual action. The establishment of RIOD is a tangible outcome of enhanced NGO cooperation.

However, one should not conclude that NGO influence on the CCD was minimal. Such an assessment would fail to take account of NGO influence over the longer term prior to the negotiations. Thus the eighth and final point here is that while the influence of NGOs during the negotiations was less than it might have been, their long-term advocacy work is at least partly responsible for the shift in thinking that distinguishes the Convention to Combat Desertification from the earlier unsuccessful international response, the Plan of Action to Combat Desertification (Chapter 2). In particular, the philosophy underpinning the convention on decentralised participation and indigenous knowledge is one that NGOs have strongly promoted to donor agencies, agricultural research organisations and other target institutions since at least the mid-1980s. This philosophy is well documented (for example, Chambers 1983; Chambers et al. 1989; Farrington 1988). International NGOs have reinforced this philosophy by publicising the achievements of NGOs and CBOs working in the field, often under difficult conditions. Hence, although the leading international NGOs behind this new way of thinking were not themselves present at the INCD, their ideas were clearly represented.

More than any previous international treaty, the CCD reflects a recognition by governments of the importance of involving local communities in designing and implementing environmental action programmes, and of the role that NGOs can play in achieving that involvement. It includes more than thirty references to the contribution of NGOs and community-based organisations. However, while their ideas helped in formulating the convention, in the long term NGOs will be judged by how well they implement it. Furthermore, and as with the negotiation process, the influence of NGOs on the implementation of the convention is probably less important than the impact of the implementation process on NGOs. As Scherl (1996) points out, the CCD has inspired NGOs to take on initiatives to a greater extent than with other conventions and it has served as a catalyst to mobilise NGOs at all levels. It remains to be seen whether NGOs can ensure that the commitments made during the negotiations are implemented, and whether these commitments can lead to long-term sustainable land management.

Notes

1 The five INCD negotiating sessions were: INCD 1, Nairobi, 24 May–3 June 1993; INCD 2, Geneva, 13–24 September 1993; INCD 3, New York, 17–28 January 1994; INCD 4, Geneva, 21–31 March 1994; and INCD 5, Paris, 6–17 June 1994. Following the conclusion of the negotiations, UN General Assembly Resolution 49/234 (December 1994) provided for the INCD to continue to meet pending sixty ratifications. The five sessions of the interim period were: INCD 6, New York, 9–18 January 1995; INCD 7, Nairobi, 7–17 August 1995; INCD 8, Geneva, 5–15 February 1996; INCD 9, New York, 3–13 September 1996; and INCD 10 (which met after the convention had entered into legal effect to address technical issues related to the first Conference of Parties), New York, 6–17 January 1997, resumed Geneva, 1 August 1997. The first Conference of Parties met in Rome, 29 September–10 October 1997.

2 Up to 1994 there were nine countries in the SADC, namely Angola, Botswana, Lesotho, Malawi, Mozambique, Swaziland, Tanzania, Zambia and Zimbabwe. Since then South Africa, the Democratic Republic of the Congo, Mauritius and the Seychelles have joined.

3 ZERO's membership now includes campaigners from SADC countries other than Zimbabwe. However, the acronym ZERO has been retained.

References

Carr, S. and Mpande, R. (1996) 'Does the definition of the issue matter? NGO influence and the International Convention to Combat Desertification in Africa', *Journal of Commonwealth and Comparative Politics* XXXIV, 1: 143–66.

Centre for our Common Future (CCF) (1993) *The Independent Sector's Network*, 22 January, Geneva: CCF.

Chambers, R. (1983) *Rural Development: Putting the Last First*, Harlow: Longman.

Chambers, R., Pacey, A. and Thrupp, L. (eds) (1989) *Farmer First: Farmer Innovation and Agricultural Research*, London: IT Publications.

Charlton, R. and May, R. (1995) 'NGOs, politics, projects and probity: a policy implementation perspective', *Third World Quarterly*, 16, 2: 237–55.

Chasek, P. (1994) 'From Stockholm to Rio: an analysis of 20 years of multilateral environmental negotiation in the United Nations system', Ph.D. thesis, Johns Hopkins University, Washington DC.

FAO (1993) 'Sustainable development of drylands and combating desertification', *FAO Position Paper*, Rome: Food and Agriculture Organisation.

Farrington, J. (1988) 'Farmer participatory research: editorial introduction' (Farming Systems Series – Number 10), *Experimental Agriculture* 24: 269–79.

Federal Environmental Protection Agency (FEPA) (1989) National Policy on the Environment, Lagos: FEPA.

Federal Republic of Nigeria (FRN) (1996) *Country Strategy Note*, Abuja: FRN.

Finger, M. (1994) 'Environmental NGOs and the UNCED process', in T. Princen and M. Finger (eds) *Environmental NGOs in World Politics*, New York: Routledge.

Grindle, M. (ed.) (1980) *Politics and Policy Implementation in the Third World*, Princeton NJ: Princeton University Press.

Haufler, V. (1993) 'Crossing the boundary between public and private: international regimes and non-state actors', in V. Rittberger (ed.) *Regime Theory and International Relations*, Oxford: Clarendon Press.

Holsti, K.J. (1977) *International Politics: A Framework for Analysis*, Englewood Cliffs, NJ: Prentice-Hall.

Humphreys, D. (1996) *Forest Politics: The Evolution of International Cooperation*, London: Earthscan.

International Institute for Sustainable Development (IISD) (1993) *Earth Negotiations Bulletin* 4, 22 (Winnipeg: IISD).

Lean, G. (1995) *Down to Earth: A Simplified Guide to the Convention to Combat Desertification, Why It is Necessary and What is Important and Different About It*, Geneva: Centre for Our Common Future/Interim Secretariat for the Convention to Combat Desertification.

Porter, G. and Brown, J. (1991) *Global Environmental Politics*, Boulder, CO: Westview Press.

Scherl, L. (1996) *Relationships and Partnerships among NGOs, CBOs and Indigenous Groups in the Context of the Convention to Combat Desertification and Drought*, Nairobi: Environment Liaison Centre International.

Sjöstedt, G. (ed.) (1993) *International Environmental Negotiation*, California: Sage.

Toulmin, C. (1993) *Combating Desertification: Setting the Agenda for a Global Convention*, London: International Institute for Environment and Development.

Toulmin, C. (1997) 'The Desertification Convention', in F. Dodds (ed.) *The Way Forward: Beyond Agenda 21*, London: Earthscan.

United Nations Development Programme (UNDP) (1995) *Building Partnerships to Support Implementation of the Convention on Desertification at National Level*, New York: UNSO/UNDP.

United Nations (1992a) *Convention on Climate Change*, New York: United Nations.

United Nations (1992b) *Rio Declaration on Environment and Development*, New York: United Nations.

United Nations (1994) *Convention to Combat Desertification*, New York: United Nations.

Willetts, P. (1996) 'Social Movements, NGOs and the Impact of the Earth Summit on the United Nations System', *Review of International Studies* 22, 1: 57-80.

World Bank (1995) *Desertification – Implementing the Convention: A World Bank View*, Washington DC: The World Bank.

4 Challenging maize

The role of NGOs in promoting the use of small grains in Zimbabwe's semi-arid regions

Derek Gunby

Maize is the dominant staple crop in Zimbabwe, and even in the semi-arid zones where it regularly fails due to recurrent droughts or inappropriate rainfall patterns it is widely grown. Over the past ten years or so a number of NGOs in Zimbabwe have sought to promote the use of traditional, drought tolerant, small grains in dry areas as a means of providing increased food security and reducing land degradation. They have approached this task in a variety of ways. As a broad approach there has been a campaign to try and get communal land farmers to replace maize with grains such as millet and sorghum. Some agencies have sought to improve the yields of small grains through the development of new hybrids. Yet others have acted to retrieve as many of the traditional varieties as possible and make these more easily available. There have also been efforts to make the processing of what is a very hard grain that much easier through the development of a de-hulling machine. It is difficult to evaluate the success of the NGOs in this work, especially without a detailed research study. On broad indices such as sales of grains, amount of land devoted to small grains production, the improved yields of small grains, or the numbers of people in dry areas who remain reliant on food relief or grain loans one would have to judge that the campaigns have not yet succeeded. There have, however, been some successes, although given the entrenched position of maize the process of change to small grains will inevitably be slow.

This chapter begins by sketching the case for increased small grain production in Zimbabwe's semi-arid regions. This is followed by a description and analysis of the rise of maize and its subsequent dominance as an arable crop. It then considers the campaigns of various NGOs in the research and promotion of small grains. The final sections analyse the strategies of the various campaigns and evaluate the effectiveness of NGOs and other agencies in promoting small grains production. It should be stressed that this chapter is not based on substantive field research but on a survey of publications and general literature in the area and on a number of discussions held with relevant NGOs and other involved agencies. It is intended as a contribution to a debate about the role of

NGOs in impacting upon national and local policy rather than as a substantive enquiry into the spread or otherwise of the use of small grains in the semi-arid areas of Zimbabwe.

The case for small grains

Small grains in Zimbabwe comprise mainly sorghum, both red and white varieties, and millet, both pearl millet and bulrush varieties. Red sorghum is most frequently consumed as beer, whilst the others are either made into a porridge meal, known as *sadza*, or used as a cattle feed. White sorghum can also be used in making bread. Commercial farmers tend to grow sorghums for the brewing and bread-making industries. Communal land farmers tend to retain most of their crops for home consumption and stock feed (Agritex Training Branch 1993).

Simply stated, the case for promoting the use of small grains in Zimbabwe's semi-arid regions is to provide improved food security in these areas and reduce land degradation. Small grains are more drought tolerant than maize and may be used both for human consumption and as a cattle feed. Because they survive in semi-arid conditions, small grains provide protection for the soil, where a failed maize crop exposes soils to erosion. There is a further argument regarding the need to maintain biodiversity. Stocks of many traditional small grains, or land races, have become severely depleted to the extent that some are in danger of extinction. Indeed some varieties may already be extinct. The reduction of the genetic pool of grains reduces future options for research and may prevent valuable new advances in crop development in the future. Nor is this a specifically Zimbabwean problem. The importance of sorghums and millets in most developing countries in the semi-arid tropics is not fully appreciated and there is a neglect of coarse grain research and development work, particularly in extension activities (Mushita 1992).

The scale of the need for redirecting crop production and research in Zimbabwe's semi-arid lands may be gauged by the fact that these dry areas occupy a major portion of communal land in Zimbabwe and are home to a large number of households. Three-quarters of all communal land is situated in areas which are unsuitable for growing maize. These areas are described as ecological Natural Regions IV and V (Vincent and Thomas 1962) and receive, on average, 450–650 mm of rain per annum in the case of Natural Region IV and under 450 mm in the case of Natural Region V. These regions cover much of the west and southern parts of the country, especially in Matabeleland North and South and Masvingo Provinces.

The areas are prone to periodic droughts and severe dry spells in the rainy season. Strictly speaking even Natural Region III is not recommended by agronomists for maize growing. If this region were added into the picture then approximately 90 per cent of all communal land farmers should not grow maize as their staple crop product. However, recent research has led to the introduction of more drought-tolerant maize varieties so that Natural Region III is not a prime target for promoting small grains. Despite the unfavourable ecological conditions many farmers in these semi-arid regions continue to grow maize or are heavily reliant upon it (ENDA-ZW 1995). The extent to which small grains are the main staple crop in Zimbabwe's semi-arid regions varies considerably. In some areas of Matabeleland South and North significant areas are devoted to small grains, but even in these more traditional-based areas there are many farmers who continue to try and grow maize, without much success. In the drier areas of Mashonaland and Masvingo maize usually predominates. Even where small grains are extensively grown yields are poor and deprivation from the seed-eating quelea birds can devastate crops. The need, therefore, is not simply to persuade farmers to grow more small grains, but also to ensure higher yields for small grains and protection against the quelea birds.

Whilst maize growing is the preferred crop for large numbers of communal farmers, there are a number of views, some of which are conflicting, on why this might be so. Some have pointed to the difficulty of processing small grain and the hard labour involved in pounding the grain into a flour (ENDA-ZW 1995). Others have argued that maize dominates partly due to a deliberate promotional strategy by seed companies and agribusiness who have considerable investments in maintaining maize production (Mushita 1992). In this argument the maize interest, which was fostered by early settlers, is seen as preventing other options for communal farmers. It is also pointed out that it is now difficult to obtain small grain seed and that the market conditions in terms of price structuring do not favour its production (ENDA-ZW 1991). Finally there is a view that younger generations brought up on the lighter, smoother texture of maize *sadza* actually prefer it and reject the more gritty *sadza* produced from millets and sorghums. This may be particularly so in the urban areas and may constitute a consumer resistance to the marketing of coarser grains. Some of these issues will be re-visited at the end of the chapter.

The dominance of maize

In many ways the spread of maize as the dominant staple crop of Zimbabwe may be seen as a metaphor for the spread of colonialism itself. Maize is not an indigenous crop and may have been initially introduced to the eastern parts of Zimbabwe through trade connections with the Portuguese in the early nineteenth century. In any event it was not widely grown prior to the colonisation of the country by the British in the 1890s. The staple crops of the Shona-speaking peoples were finger millet (*ropoko*) and bulrush millet (*mhunga*), whilst for the Ndebele people in the west the principal crops were sorghum (*amabele*) and bulrush millet (*inyauti*) (Illife 1990). These traditional crops, or land races, had many varieties and must have been developed over the previous thousand years or so, following the arrival of the earliest Bantu farmers.

Whilst the traditional land races are undoubtedly much more adapted to local dry conditions than maize, even these crops can and sometimes do fail in the face of severe droughts. There are some recorded instances of severe drought and famine in Zimbabwe prior to colonisation and the widespread adoption of maize. The 1790s in the Manyika region, 1827–9 and 1833–6 are some of the more serious instances (Illife 1990).

Maize has many important attributes as a crop. It is capable of much higher yields than traditional local crops (although this tendency has been enhanced through massive research and experimentation over many decades), it is easier to process into a flour for consumption as a porridge or 'meal', it can be used in its 'green' state, and it is not prone to the predations of grain-eating birds, of which the most common are the quelea birds, dubbed elsewhere as 'the feathered locusts' (Mundy and Jarvis 1989). For all these reasons maize is much more attractive both as a staple crop and as a cash crop and the early settlers soon moved into substantial maize production.

The process whereby maize supplanted small grains in practically all regions and localities of Zimbabwe as the primary food staple has not been charted in detail to the author's knowledge. However, it is clear that two broad processes were at work. The first set of mechanisms related to what may be called the economics of grain production, whereby the buying and selling of grain moved from a pre-colonial system of bartering to a post-colonial one of marketing arrangements and regulations which instituted both state control of buying and selling and private commercial farming dominance. These moves paralleled the seizure by the settler regime of large tracts of the most fertile land and the enforced movement of local peasant farmers to the more marginal lands (Moyana 1984). The new colonial, commercially-oriented, arable farmers concentrated heavily on maize as a cash crop, both for internal consumption and export. Agricultural services were geared to support and advice on maize production and research stations were oriented to experimenting into various methods of improving the yields, consistency and pest removal for maize. As for the other grains, only in sorghum, which is an important ingredient of local beer, was there any significant investment of expertise and effort.

In terms of marketing arrangements, state regulations were established to regulate buying and selling of grains and private grain companies set the economic value of grain. Both aspects reinforced maize production. In 1931 the Southern Rhodesian government introduced the Maize Control Act which established a Maize Control Board to monopolise the purchase and sale of maize in Mashonaland and later the whole country (Illife 1990). The effects of this Act and its subsequent amendment in 1934 were not only seen to disadvantage the African farmer seriously, but also to strengthen the hold of the European commercial farmer on grain production and marketing (Moyana 1984). Despite criticisms levelled at the protection afforded to the European farmers, which hampered African farmers selling both maize and especially small grains, these measures persisted.

The second set of mechanisms whereby maize came to dominate over traditional land races lies in the area of what may be termed 'preference through habit'. In other words maize became a preferred product for the vast majority of local people on the basis of taste, texture and ease of preparation. Initially there may well have been resistance on the grounds that maize meal had a different taste and texture to millet or sorghum meal, but this is not documented. What is apparent now is that very few local people, especially in urban areas, are said to actually prefer the coarser meals produced from small grains, despite the fact that they are probably healthier, containing as they do more roughage. The preference for refined maize meal, which produces a white, smooth flour, may be likened to the introduction of white bread in Europe and America in the early twentieth century, both having been accompanied by massive and sustained advertising campaigns.

The country's fixation with maize has continued to the present day, although there has been a growing awareness over the past ten years or so, in certain quarters, of the need to address the problem of food security in the dry semi-arid regions of the country through an alternative grain policy. It is perhaps surprising that even during the first few years of independence so little attention was given to small grain production in the drier areas. It was less surprising that throughout the long years of colonial rule there had been no research into the needs of the peasant farmer generally (Chavunduka 1982) and no attention to small grains. The government's research efforts had all been directed to the needs of the commercial farmer. In many ways such interests determined that there should be a system of dependency of local people upon the commercial sector and maize. To have developed farming methods in the Tribal Trust Lands (as the communal lands areas were referred to prior to 1980) that might lead to self-sufficiency or even grain surpluses would have posed a threat to the profitability of the maize-growing commercial farmers.

After independence, although a new Agricultural Extension Service (Agritex) was established to replace the two separate extension services, Conex (Department of Conservation and Extension) and Devag (Department of Development and Agriculture), which operated in the commercial and communal farming areas respectively, the emphasis on crop production for all areas remained based on maize production. In part this was due to a sudden and encouraging expansion of maize production by communal land farmers between 1980 and 1982. The Chavunduka Report on the Agricultural Industry in 1982, whilst deploring the lack of attention in past agricultural research and advice services to the needs of communal land farmers and noting the inappropriate use of maize in the drier areas, did not suggest promoting small grains or make any specific recommendations related to small grains (Chavunduka 1982). Nor was there any major policy shift by the government until well into the 1980s. Even as late as 1993 a publication by the Office of the Vice-President called *Zimbabwe: The Drought Relief and Recovery Programme 1992/93* concluded with a photograph of high-standing maize and the caption 'Life returns to normal: The good rains last season brought with it a bumper harvest after farmers were given free seed packs by the government' (Government of Zimbabwe 1993).

The major seed companies in Zimbabwe continue to promote maize in all regions, despite the fact that it is often unsuitable in the drier areas. The companies are able to promote maize production in the semi-arid communal lands by advertising, extension services and the selection of model farmers who are given support in growing certain varieties of hybrid maize. At the end of each season seed companies proudly advertise the 'winners' of their maize yields competitions, featuring communal land farmers in a variety of settings including the semi-arid regions. Agritex extension officers are also often targeted by seed companies to promote various maize varieties (Mushita 1992).

The culture of maize can therefore be seen to be pervasive, despite successive droughts in recent years. Most major agricultural agencies continue to espouse maize and to research new high-yielding hybrids. The seed companies and the Grain Marketing Board continue to be biased in favour of maize. But there has been some change. Agritex, in particular, is now much more active in promoting small grains in semi-arid regions and often links up with NGOs. Some agriculture research stations are also involved in small grains research. These changes have been occurring slowly since the mid-1980s. The role of NGOs and international agricultural agencies has been influential in this process. The next section describes the activities and processes of these change agents.

The promotion of small grains

It would appear that 1983 was a pivotal year in the development of a more concerted effort to promote the use of small grains in Zimbabwe's semi-arid regions. During that year several new initiatives emerged from both local and international NGOs that sparked a more energetic effort. The common factor was undoubtedly the experience of the 1982/3 drought which jolted the whole country out of the euphoria of the previous two good rainy seasons when there had been a massive expansion of maize growing in the communal lands.

In 1983 the annual general meeting of the Organisation of Rural Associations for Progress (ORAP) adopted a four-point strategy for enhanced food security which gave priority to a recourse to traditional seeds (Kempadoo and ORAP 1991). In the same year Environment and

Development Activities (ENDA), an international NGO with its headquarters in Dakar, Senegal, established a branch in Zimbabwe and immediately commenced a survey to identify cultivars[1] sown in semi-arid areas and how they are preserved and used. From this point on activities began to multiply and more NGOs began to develop programmes aimed at encouraging small seeds production.

In 1987 a number of NGOs active in the area of small grains promotion came together to form the Zimbabwe Seeds Action Network (ZSAN). The members included ORAP, Silveira House, the Manicaland Development Association (MDA) and ENDA, which acted as the secretariat. Support was obtained from a Dutch donor, the Netherlands Organisation for International Development Cooperation (NOVIB), which provided funds for a four-year exercise between 1988 and 1991. The central objective of the network was to ensure the supply of indigenous crop seeds to communal households in the drought-prone regions of Zimbabwe. Whilst a national secretariat undertook coordinating and networking activities, each participating organisation was left to devise its own programmes within the broad objective. ENDA, with larger resources at its command, was able to undertake a major multifaceted programme, which it called the Indigenous Seeds Project. This project was also funded by NOVIB and ran for an initial four years from 1987/8 to 1990/1.

Shortly after independence in 1983, the International Crops Research Institute for the Semi-Arid Tropics (ICRISAT) established itself in Zimbabwe as part of a Southern African Development Community (SADC) initiative, basing itself at the government's agricultural research station at Matopos in Matabeleland. In the mid-1980s it agreed to collaborate with ZSAN on further germ plasm[2] evaluation, seed and information exchange. ICRISAT also began a series of trials related to sorghum and millet hybrids as part of a Sorghum and Millet Improvement Programme (SMIP).

The period of the 1990s was been dominated by two further serious droughts, in 1991/2 and 1994/5. The 1991/2 drought was probably the worst in Zimbabwe for about a century, both in terms of the degree to which practically all parts of the country were affected and for the lack of rain. The 1994/5 drought was almost as severe, and in some regions was worse than 1991/2, but it was less widespread. The generalised failure of the various actions to promote small grains in the semi-arid regions was manifest in the almost complete inability of communal land farmers to secure any harvest during these dry years and the consequent need for massive assistance in terms of food assistance and feeding programmes. The experience of these droughts has led to additional NGO activity in the promotion of small grains. In Matabeleland, Christian Care and the Lutheran World Federation are both now active in working with local farmer groups to promote small grains production. The Dutch international NGO, the Netherlands Development Organisation (SNV), has also assisted in small grains work with local farmer groups. The local provincial offices of Agritex have adopted a more aggressive attitude to small grains promotion, producing their own leaflet and linking up with local NGOs in a variety of joint activities such as field days (Agritex 1993). ZSAN suffered a collapse after the initial four-year funding in 1991 and has not really recovered. Silveira House and the MDA dropped out prior to that and ENDA went through some major organisational changes, although it was able to secure a further four-year funding programme for its small grains work from NOVIB in 1994.

The overall picture of activity at present is difficult to assess. The lack of national coordination through the collapse of ZSAN has probably reduced the impact at national and regional levels. On the other hand there is evidence of increased activity at the local level with some increased inter-agency cooperation.

Government policy on grains

Government policy has only recently recognised the need for increased research into and promotion of small grains in semi-arid regions in a clear way. Policy makers before independence completely failed to perceive that such a solution to the food security of Africans living in the Tribal Trust areas existed.

Following independence the Transitional National Development Plan stated that there was a need for research into suitable crops for arid areas (Government of Zimbabwe 1982). However, there is little indication that this was followed up in any urgent fashion. The 1982 Chavunduka Report on the Agricultural Industry failed to identify the need to research and promote drought-tolerant grains in dry areas, preferring a solution based on improved livestock-based farming. The report did say that there were almost no data available on farming conditions in communal lands and made a number of recommendations to improve the situation, but the issue of maize versus small grains did not surface (Chavunduka 1982). It is noticeable that the First Five Year National Development Plan of 1986–90 failed to repeat the recommendation for research into suitable crops for dryland areas. The section on agriculture continues to extol maize, anticipating further increases in production, and makes no mention of the impact of the droughts of the early 1980s except in terms of their impact on wheat production (Government of Zimbabwe 1986).

The impact of the 1991/2 drought may have led to the much more positive stance adopted by the government in its Second Five Year National Development Plan, 1991–95. This document clearly states: 'Research into drought resistant crops will be assigned top priority. Farmers in dry areas will be encouraged to grow early maturing food crop varieties and intercropping' (Government of Zimbabwe 1991). Whilst this position still left much to be desired by the small grains lobby it did mark a significant step forward. In any event provincial Agritex offices in the dry areas had been progressively advocating small grain production from the late-1980s onwards.

Perhaps the most visible sign that the government had seriously adopted a small grains policy came with the introduction of sorghum seeds into the 1994 free seeds packs provided to communal land farmers in Natural Regions IV and V. This approach was followed through in the distribution of grain loans, but there were complaints that these loans were insufficient.

The activities pursued by small grains promoters

The work of NGOs promoting the use of small grains since 1983 may be considered from the grassroots upwards in terms of eight strategies.

First, advocates have acted to persuade farmers to grow small grains. All of the members of ZSAN worked directly with farmer groups to promote the use of small grains. In this work appropriate seeds were provided and training offered at all stages in the production process. The idea was to demonstrate the effectiveness of small grain production in dry areas through actual examples in the field. The method of NGOs working alongside selected farmer groups had been pioneered in the western part of Zimbabwe by the Hlekweni Training Centre (de Graaf 1987). This was established as early as 1967 and many of the advocates of participatory rural and agricultural development today were involved at some stage in Hlekweni's work.

The MDA used land at Chitepo Agricultural Training Centre to illustrate to surrounding farmers the advantages of small grain production and the best methods of farming such crops. Silveira House worked in Mashonaland East province and introduced the small seeds programme into their ongoing food security projects among farmer groups. ORAP worked among farmers in both Matabeleland South and North, targeting especially the Kezi and Tsholotsho districts. With their superior resources ENDA adopted a more rigorous approach selecting five wards of thirty villages in Mutoko (Mashonaland East), Chipinge (Manicaland), Zvishavane (Midlands) and Plumtree (Matabeleland South). A ward-level seed committee was formed constituted by two representatives from each of the six villages in the ward. The Ward Seed Committee then selected participant farmers, organised meetings and assisted in coordinating local activities. Research was conducted utilising these participant groups so that the results of seed growing could be made available to a variety of research and other agricultural institutions. ENDA's approach was therefore integrated in that local farmer participation was intended to promote the use of small grains at local level and provide comprehensive, scientific data to challenge those who did not accept the value of growing small grains (ENDA-ZW 1990; Mushita 1992).

In addition to promoting the use of small grains the ZSAN group also tackled the issue of grain storage. Some surveys indicate that as much as

60 per cent of food may be lost post-harvest because of poor storage methods (ENDA-ZW, 1991). Improved granaries were built in a number of villages to demonstrate their effectiveness and to train villagers in improved methods. This type of work has been taken up in a national campaign by Agritex.

Second, NGOs have provided farmers with appropriate seed. The exploration and collection of germ plasm material was the foundation and building block of the project. Not only did the NGOs organise the collection and duplication of small grain seeds but also open-pollinated varieties of maize. In ENDA's case they collected twenty-seven maize, fifty-five sorghum and thirty-four pearl millet populations. Having collected the seeds NGOs developed a seeds multiplication programme. In addition, the national secretariat organised seed collection expeditions. ORAP used one of their farms at Figtree in Matabeleland South to produce seed. However, it is generally admitted that these efforts were insufficient to provide seed on a mass scale. One of the results of years of maize dominance has been the virtual elimination of small seeds availability. The initial ZSAN programme involved some 4,000 farmers in a seeds multiplication programme. Today it is estimated that number is down to 3,000. These programmes have helped to ensure the retention of many land races that would otherwise have been lost. ENDA has established a gene bank and information centre to continue this work. However, there is still a long way to go before it can be said that small grains are widely available and accessible to the communal land farmer.

Third, NGOs have improved the technical inputs to processing. One of the complaints from potential beneficiaries concerning the use of small grains is that they are much more difficult to grind than maize and that the usual local grinding mill cannot grind small grains. The traditional method involves many hours of pounding by women to produce a usable flour. To overcome this problem a number of agencies have been working on developing a machine that removes the hard hulls of the small grains. Work on de-hullers, as they are known, was successfully undertaken by ENDA in a project that culminated in a programme to introduce some sixty de-hullers in rural areas over four years. The ENDA de-hullers are now in commercial production by three Zimbabwean companies. ORAP also engaged an engineer to work with them on developing a suitable de-huller.

Whilst some promoters of small grains are against hybrids, because they are not capable of self-propagation, there is an advantage in that they have less tough skins. The work of ICRISAT in developing hybrid sorghums and millets may therefore overcome the need for special de-hullers, if hybrid seeds become widely available.

As mentioned previously, one of the more significant problems facing the growers of small grains is that they are susceptible to considerable loss through the activities of the quelea bird. The quelea can devastate small grain crops. Efforts to produce new techniques to scare the birds away or reduce their numbers seems to have received very little attention. There is a Problem Bird Control Unit within the Department of National Parks and Wildlife Management which locates and sprays flocks of quelea, but it is seriously underfunded. In 1996 press reports quoted the Problem Bird Control Unit as saying that quelea numbers had increased dramatically in the recent rainy season and estimated that there was a current Zimbabwe population of 500 million (*The Chronicle*, May 1996).

Fourth, NGOs have encouraged Agritex to support the campaign for small grains at provincial and local levels. In discussion with several of the relevant NGOs in small seeds promotion it became apparent that their primary strategy was to convince through example. Notions of campaigns and seeking to convince such authorities as Agritex did not figure strongly in their plans. However, it is clear, for one reason or another, that there has been an increase in inter-agency work on small grains. The Tsholotsho district in Matabeleland North provides an interesting example of cooperation between NGOs, Matopos Research Station, ICRISAT and local farmer communities in millet production. In the Khulumsenza area of Tsholotsho, farmers were provided with a new millet hybrid, called Okashana, developed by ICRISAT, and Agritex and the other agencies assisted in its distribution and held training sessions with local farmers. The resultant high yields received considerable publicity (*The Chronicle*, February 1996).

Fifth, NGOs have worked to persuade policy makers at the national level that small grains are viable and appropriate. Whilst there does not appear to have been a systematic effort by the national secretariat of ZSAN to target key government agencies in a concerted campaign, there were separate initiatives taken at national level to involve certain national bodies in collaborative work. The National Farmers Association of Zimbabwe, now the Zimbabwe Farmers Union (ZFU), which represents some 50,000 communal farmers, actually approached ZSAN and asked for small grain seeds for distribution to their members. The association also participated in collecting germ plasm. Other organisations, such as the Permaculture Association, the National Botanical Garden and ICRISAT were also involved in various aspects of germ plasm collection and in other aspects of research. The Crop Breeding Institute, which comes under the Ministry of Agriculture's Department of Research and Specialist Services, agreed on a wide range of collaborative work, including on-farm trials. ENDA-ZW's credibility with policy makers was recognised when it became a founder member of the Zimbabwe Biotechnology Advisory Committee (ZIMBAC).

Sixth, NGOs have successfully persuaded members of the agricultural research community that their trials and research on small seeds have been undertaken in a scientifically sound manner. For example, in order to gain acceptance of the results of their research within the agricultural establishment, ENDA ensured that their socio-economic surveys undertaken in the 1988–90 period were scientifically rigorous. They were able to demonstrate this to other agriculturists. In the germ plasm evaluation, 1,009 types of sorghum and 401 of pearl millet land races were characterised under research station conditions (ENDA-ZW 1990; Mushita 1992).

Seventh, NGOs have coordinated the national efforts of small grains advocates. ENDA, through its leadership of ZSAN between 1988 and 1990, played an important role in coordinating efforts, both in terms of promotion at local level and in securing germ plasm, seed multiplication and research results. Unfortunately the system began to break down before the end of 1990 with the withdrawal of Silveira House and MDA from the network. After 1990 ENDA seemed to relinquish its coordinating role and ZSAN ceased to operate. In 1995 there appears to have been a revival in ENDA's interest and activity in promoting small grains. In October 1995 it published, with the support of NOVIB, a broadsheet entitled *Seeds Action Briefing* (ENDA-ZM 1995). This publication received a certain amount of national press coverage and led to a higher profile of small seeds work in 1996. However, national coordination of NGO and other potential organisations in the crop production field has not yet emerged.

Finally, the NGO campaign in Zimbabwe has had an impact at the international level. In the late 1980s and early 1990s ENDA played a central role in the formation of a SADC Food Security Network and acted as its secretariat. ENDA was able to involve ZSAN in a number of regional workshops and meetings. ENDA was also involved in the Southern African Network of Development Organisations (SANDON) in cooperation with NOVIB (Mushita 1992).

Evaluating the impact of small grains advocacy

Little effort appears to have been made by NGOs to influence the Grain Marketing Board (GMB) and major grain-buying concerns to give more favourable consideration to small grains. In fact given the prevailing conditions prior to the Economic Structural Adjustment Programme (ESAP) there was limited space for NGOs to manoeuvre. The GMB was forced to buy all the small grain on the market. Because demand was low it built up substantial stocks. Since deregulation in 1991 the GMB is free to buy as much grain as it wants and to set the selling price at market levels. This has led the GMB to buy much less and to reduce the price it pays for small grains. Under these conditions there is even less that NGOs can do to influence the GMB, which now operates solely on market criteria.

Since deregulation, private milling companies and others are able to buy grain direct from the producer. The milling companies buy white sorghum, which is used in making bread, and the brewers buy red sorghum, which is used in beer making. There appears to be no major interest in buying millet. The GMB is not convinced that there is likely to be a growing market for small grains. However, GMB's monopoly position has been ended and there are plans through ENDA-ZM to establish medium-sized commercial small-grain milling companies to produce different brands of small-grains meal for urban and possibly export consumption (*The Sunday Mail*, March 1996). A local Bulawayo company, Induna, has already charted a path in this direction and this may presage a general improvement in commercial outlets for small grains producers.

It is not easy to measure the impact of the small grains advocacy work on farming practice. It is the view of some of the NGOs involved that there has been an increase in the amount of land given over to small grains, but even if this were so it is likely to be only a marginal increase. The general evidence is that, despite numerous severe droughts in the last fifteen years, the overwhelming majority of farmers in semi-arid regions still grow considerable quantities of maize. In Matabeleland North, for instance, in 1996 Agritex reported that communal farmers had a total of 69,421 hectares under maize, compared to 32,810 hectares under sorghum and 29,989 hectares under millet.

On the other hand, it has to be accepted that in order to achieve a major shift in farming practice a large number of factors need to be put into place, including knowledge, opportunity, preference and performance. There has probably been a significant increase in knowledge about small grains among farming communities in the semi-arid regions, if only because a significant number of NGOs and, more recently Agritex extension officers, have been engaged in disseminating such information over several years. Opportunities for growing small grains have remained limited due to the failure of seed companies to develop and market such grains and to the loss of local indigenous grain stocks. ZSAN sought to compensate for this problem by arranging for local seed multiplication programmes. Whilst this must have had an impact it was probably only a marginal one. A more concerted national effort would be required to ensure that every farmer who wanted to switch from maize to small grains had easy access to the grain at the right time and at the right price.

It is unclear from published sources and general comment whether, from the point of view of taste, texture and nutritional value, local people prefer maize to millet or sorghum. Advocates of maize tend to emphasise that their product is preferred and that efforts to introduce small grains will meet consumer resistance. It will, however, come as no surprise that advocates of small grains claim that local villagers prefer the healthier and stronger tasting millets and sorghums.

The performance of small grains in terms of yields still mitigates against their wider adoption. Although maize often fails in the semi-arid areas, when it does succeed it always yields more than small grains. In the Matabeleland example mentioned above, the yields for maize were 1.09 tonnes per hectare, for sorghum 0.75 tonnes per hectare, and for millet 0.82 tonnes per hectare. The current work of agencies such as ICRISAT in developing higher-yielding varieties may be an important factor in encouraging farmers to increase their small grain production.

Conclusions

The broad conclusions of this study may be stated as follows.

First, local NGOs have sought to influence policy change through grass-roots activities designed to create a demand among the farmers for small grains. In this there have been localised successes, but when looked at in overall terms only marginal progress has been made. In terms of the four NGO approaches mentioned in Chapter 1 – complementary activities, consciousness-raising, confrontation and collaboration – the approach of the NGOs involved in the campaign to promote small grains in Zimbabwe's semi-arid regions has essentially been complementary and, more recently, collaborative. The NGOs have also played an important consciousness-raising role among farmers. Confrontational tactics have been considered counter-productive and have been avoided.

Second, international NGOs such as ENDA, while seeking to influence local opinion also sought to achieve wider policy shifts by influencing opinion within the agricultural research community. In this they may have been successful, although again it is difficult to verify. Their work has undoubtedly added to understanding and knowledge about small grains, and their new hybrid strains may assist in overcoming consumer resistance. However, recent work in developing drought-tolerant maize hybrids may provide an easier path to food security in the semi-arid areas since it will not have to overcome the many obstacles that small grains face.

Third, maize production and consumption remains deeply entrenched and, whilst the work of the small grains advocates may have dented this culture, it has yet seriously to challenge it.

Fourth, national policy changes have occurred, as witnessed in the clear statement in the 1991 Second Five Year National Development Plan and perhaps more significantly by the inclusion of small grains in the seed packs given to farmers in Natural Regions IV and V during 1993, but there has not been a comprehensive action plan by government. There has been more rhetoric in favour of small grains, especially after the two terrible drought years, but much more can and needs to be done if small grains are to be widely adopted as an integral part of a national food security plan.

Fifth, the slowness of the government in adopting a clear stance advocating small grains and following it through with action plans is claimed by some to be due to the fact that the issue is confined to certain provinces. There may be some validity in this argument, but the numbers of people who live in Natural Regions IV and V are considerable and the burden on the state of providing food relief and grain loans is enormous. A more likely explanation is that there is a lack of agreement within either the Agriculture Ministry or the various component parts of the agriculture industry on the desirability or feasibility of a concerted campaign to shift crop production in semi-arid regions from maize to sorghum and millet. In the words of one of the local activists in a dry region: 'As long as we keep getting the odd good rainy season we will remain a predominantly maize growing area.'

Notes

1 A cultivar is a plant species produced from natural species and maintained by cultivation.
2 Germ plasm is the sum total of hereditary materials in a species.

References

Agritex Training Branch (1993) *Grow Small Grains*, Zimbabwe: Agritex.

Auret, D. (1990) *A Decade of Development in Zimbabwe 1980–1990*, Zimbabwe: Mambo.

Chavunduka, G.L. (1982) *Report of the Commission of Inquiry into the Agricultural Industry, Chairman: Professor G.L. Chavunduka*, Harare: Government of Zimbabwe.

ENDA-ZW (1990) *Zimbabwe Seeds Action Network: Second Annual Report*, Harare: ENDA-Zimbabwe.

ENDA-ZW (1991) *Zimbabwe NGOs National Food Security Workshop*, Harare: ENDA-Zimbabwe.

ENDA-ZW (1995) *Seeds Action Briefing* 1, October.

Government of Zimbabwe (1962) *Report of the Commission of Inquiry into Maize and Small Grains Industry*, Harare: Government of Zimbabwe.

Government of Zimbabwe (1982) *Transitional National Development Plan*, Harare: Government of Zimbabwe.

Government of Zimbabwe (1986) *First Five Year National Development Plan*, Harare: Government of Zimbabwe.

Government of Zimbabwe (1991) *Second Five Year National Development Plan*, Harare: Government of Zimbabwe.

Government of Zimbabwe (1993) *The Drought Relief and Recovery Programme, 1992/93*, Harare: Office of the Vice President.

Graaf, M. de (ed.) (1987) *The Importance of People: Experiences, Lessons and Ideas on Rural Development in Zimbabwe, Hlekweni and Beyond*, Bulawayo: Hlekweni FRSC.

Heinrich, G. and Mangombe, N. (1995) *Collaborative On-Farm Research on Sorghum and Pearl Millet in Zimbabwe: Review of the 1992/93 Trials*, Zimbabwe: ICRISAT Working Paper 95/01.

Iliffe, J. (1990) *Famine in Zimbabwe*, Harare: Mambo.

Kempadoo, P. and ORAP (1991) *Zenzele, The ORAP Way*, Bulawayo: RCP.

Moyana, Henry (1984) *The Political Economy of Land in Zimbabwe*, Zimbabwe: Mambo.

Mundy, P.J. and Jarvis, M.J.F. (1989) *Africa's Feathered Locust*, Zimbabwe: Baobab.

Mushita, T.A. (1992) *Zimbabwe Seeds Action Network Four Year Report*, Zimbabwe: ZSAN.

Vincent, V. and Thomas, R. (1962) *An Agricultural Survey of Southern Rhodesia: Part 1; Agro-ecological Survey*, Salisbury: Ministry of Agriculture.

5 The campaign for water from the Zambezi for Bulawayo

Derek Gunby, Roger Mpande and Alan Thomas

> 'PLEA OF A DYING CITY'
> 'Over 3,000,000 people in Bulawayo and Matabeleland may be relocated to other parts of the country as a result of the drastic water situation.'

This dramatic text is from the first advertisement placed by the Matabeleland Zambezi Water Project (MZWP) in October 1991, shortly after its formation. The frightening prophecy of mass evacuation was only avoided when, with twenty-two days water supply available, heavy rains rejuvenated the city's dams in late 1992.

The MZWP was a non-governmental organisation (NGO) set up to work for a permanent solution to Bulawayo's water supply problems by bringing water from the Zambezi River. It was argued that this would benefit the whole of Matabeleland by releasing existing city dams for irrigation projects in Matabeleland South and providing water for irrigation along the pipeline route in Matabeleland North.

After an energetic campaign with strong populist overtones by MZWP, the Zimbabwean government adopted the Zambezi pipeline idea in principle by writing it into its Second National Five-Year Development Plan and supporting a two-stage feasibility study, which commenced at the end of 1994. At the same time leading government and ruling party figures established a new NGO, Matabeleland Zambezi Water Trust (MZWT), with representation from all sectors. The preliminary study was jointly approved by the government and MZWT in December 1995. The final feasibility study and an Environmental Impact Assessment were presented in September 1996, recommending a staged approach with a series of water supply improvements before constructing a dam at Gwayi-Shangani, which could later be linked to the Zambezi.

Thus the campaign has succeeded in bringing the idea of tapping the Zambezi into the national water planning process, and there is now clearly a role in this process for NGOs, the private sector, communities, and local and provincial as well as national government. On the other hand, several issues are unresolved. The campaign notably failed to force a specific commitment on a date for completing the link to the Zambezi, and was unable to maintain a role for MZWP itself in building or managing this link.

This chapter looks critically at why the government changed its policy and at the outstanding unresolved issues. The next section outlines the water situation in Matabeleland. Then we relate the story of the campaign and analyse the factors explaining both the policy influence of NGOs such as MZWP and the limits to their influence.

Water shortage and the history of water development in Matabeleland

Matabeleland South and North are two provinces located in western Zimbabwe, part of the dry Kalahari-Namib region. Bulawayo, Zimbabwe's second city and the largest in Matabeleland, has suffered frequent water shortages. Initially, wells provided water to Bulawayo. However, a more reliable and convenient water supply was required as the population rapidly grew.

Bulawayo's water supply history shows a systematic development programme from 1928 to 1976, when the last major water development scheme was undertaken. With one exception, dams were situated in the southern catchment area, beyond the watershed that runs close to Esigodini, some 40–50 km south-east of the city. During this period the time between the completion of a major water supply scheme and the establishment of a new one averaged eight years. However, from 1976 to 1996 no major new water scheme was developed.

Since 1980, Bulawayo has introduced water rationing and restrictions at least six times. In 1992, at the height of the worst drought year in living memory, Bulawayo was declared a Public and Underground Water Shortage Area. Although domestic water rationing is normally introduced first, the impact of full water rationing is most heavily felt in the industrial sector which uses 37 per cent of water consumed in the city. In the drought of 1991–2, a number of companies reduced their activities to a minimum, retrenching workers or working a shorter week. At least two companies relocated altogether.

Another adverse impact of water rationing was the resultant deterioration in the health sector. By May 1991 the reduced water use in Bulawayo due to rationing had resulted in an increase in the number of sewer blockages. There was also an increase in random and illegal drilling of boreholes. Food hygiene, sewerage disposal and, to some extent, personal hygiene were the main areas of concern as supplies dwindled. Although there was no apparent lowering of food hygiene standards, the City Department of Health envisaged that, as water became scarce, some food business operators might resort to the use of water from alternative sources, some of which might be polluted.

Water shortage has also been a cause of land degradation throughout Matabeleland. Although the current population growth of Matabeleland region is relatively low for Africa (2.8 per cent per annum), possibly due to out-migration as a response to drought, it is sufficient to cause an increase in demand for crop and grazing land. This entails clearing approximately 10,000 hectares of forests every year to make way for subsistence agriculture.

With the unreliability of rain in the region, the expansion of agricultural land has not resulted in a related increase in crop production. Thus, as long as no permanent sustainable resource strategy is identified that emphasises minimal natural resource destruction, the region will continue to be degraded until the damage to the environment may become irreversible. Current estimates indicate that on average Zimbabwe loses about 50 tonnes of soil per annum on cultivated lands. Most of this degradation occurs along major river systems as these are the only areas where the local population can have access to water for both domestic and livestock use.

Land degradation is most clearly visible in the most densely populated parts of the region, that is, the communal lands. This points to a correlation between fragile, sandy soils, population densities and water shortage. In these areas large tracts of land have been deforested and are subject to accelerated water and wind erosion. Apart from the long-term negative effect on the potential productivity of the land, the loss of soil results in the siltation of rivers and dams. Over the past twenty-five years virtually all the rivers of the area have become seasonal in their flow, whereas previously many of them used to run perennially.

Drought has become almost a permanent feature. Six drought seasons with variable effects have occurred in Matabeleland in the past twenty years. Thousands of cattle have died and crops failed. The government's expenditure on drought relief has increased, with substantial further amounts disbursed by NGOs, as the water shortage has become a major concern to development agencies working in the region. Long-term agricultural strategies need to be identified, and two options are the cultivation of drought tolerant crops and irrigation. Sourcing water from the Zambezi could potentially provide the latter as well as solving Bulawayo's water crisis.

The story of the campaign

The dream of the Zambezi pipeline

The River Zambezi flows approximately west to east some 450 km north of Bulawayo. Following the first recorded proposal to build a pipeline from the River Zambezi to Bulawayo in 1912, the idea was raised with the colonial administration on at least six separate occasions (Operation Pipeline 1991). Shortly after independence the City Council again raised the possibility. A report was subsequently compiled in 1983 by an Italian engineering company, concluding that the project was financially viable if linked to large enough irrigation schemes.

This proposal may have failed partly because of the armed struggle between the government of the Zimbabwe African National Union (ZANU) and 'dissidents' from the opposition Zimbabwe African People's Union (ZAPU). This low-grade, but bitter war between a predominantly Shona-led government and representatives of the Ndebele, the majority ethnic group in Matabeleland, severely disrupted development in Matabeleland and the Midlands Provinces until 1988. In that year, the Unity Accord was signed between the government and ZAPU, leading through what Stoneman (1992) has called a 'shotgun merger' to a united party (ZANU-PF) and the incorporation of ZAPU leaders into government posts. Joshua Nkomo, the leader of ZAPU and the foremost Ndebele spokesman, was made Vice-President.

The pipeline idea was again raised in 1988 when an engineer at the University of Zimbabwe wrote a paper examining costs in some detail (Mitchell 1988). The following year two MPs pressed the government on the issue (Pearl 1990), but there was still no progress. The government's position over many decades had been consistent: the project costs were too high and alternative arrangements had to be made. But no alternative provision had been put in place by 1991 when the century's most devastating drought threatened Zimbabwe's second city.

The birth of the Matabeleland Zambezi Water Project (MZWP) as an NGO

The increasingly parlous state of Bulawayo's water supply became apparent after poor rains in the 1990–1 rainy season. Water rationing was again imposed in January 1991. The city's dams were below 30 per cent full and the high rate of evaporation in the dry season coupled with continued normal consumption could, if the 1991–2 rainy season also failed, produce a crisis situation in late 1992.

By March 1991 various individuals were publicly espousing the Zambezi pipeline idea, including Eric Bloch, a prominent economist and broadcaster, Alderman Mike Constandinos, who had long campaigned on the issue, and John Pierson, who had been Bulawayo's City Engineer in the 1950s. The Sandwich Club, an informal civic group of Bulawayo professionals, businessmen and industrialists, asked Pierson to prepare a 'fact paper' on the feasibility of the Zambezi pipeline and called upon the government to implement the project. After failing to persuade Joshua Nkomo to take the paper to government, the club asked Dumiso Dabengwa, then Deputy Minister of Home Affairs and arguably the most popular local political leader, who accepted the mission. Later Nkomo met the Sandwich Club with Jevan Maseko, Governor of Matabeleland North (the highest position in Provincial Government, with Ministerial status) (Dube 1994), but this initiative did not make progress. The Vice-President certainly viewed the Sandwich Club with suspicion and the agitation for the Zambezi pipeline may have been seen in some quarters as essentially white-led. However, this is speculative.

Alderman Constandinos was prepared to take the Zambezi pipeline issue to the City Council on a formal basis, but felt that additional technical information would assist. This led, through Pierson's contacts in South Africa, to a report prepared by an engineering company, Watermeyer, Legge, Piesold and Uhlman (WLPU) and paid for by South African pipe companies, presumably in the hope of future business (WLPU 1991). This report helped increase support for the proposed project. Bulawayo City Council backed it, but recognised the need for a wider group with strong political weight to lobby central government.

MZWP became that wider group. The precise events surrounding its formation are unclear. Several individuals claim to have initiated the idea. Whatever the sequence of events, a number of concerned local figures came together and formed MZWP as a trust in October 1991. The six trustees represented different sectors of the community, ensuring that MZWP had a fairly broad base. The co-chairmen were Dabengwa and Maseko, and the other four trustees were Constandinos and three leaders of the white business community. Thereafter, MZWP became the spearhead of a drive to gather local money and support for the Zambezi pipeline. It adopted a twin strategy of appealing to local pride and self-help action as well as striving to influence national and international opinion.

The concept

The Zambezi River has vast quantities of water, with average low flows exceeding 15 million cubic meters per day. Zimbabwe's main usage of the Zambezi is for hydro-electric power generation at Kariba, while only slightly over 1 per cent of the minimum flow is abstracted. This is mainly for the urban centres of Victoria Falls, Hwange and Kariba, including industry and mining in Hwange. Bulawayo's current peak demand would require little more than another 1 per cent of the minimum flow.

The WLPU report considered two abstraction points (Figure 5.1). One was at Deka where Hwange Colliery and Hwange Power Station already have their pumping stations. The second, at Victoria Falls, would make the pipeline 84 km longer, but is 500 metres above the first and so would incur lower pumping costs. It could also serve Hwange North area with possible water for irrigation. The number of pumping stations varies according to the pipe diameter, quantity of water taken and the route chosen, but WLPU's preferred option of Deka would have required one main station and six booster stations.

Figure 5.1 Alternative routes for the Matabeleland Zambezi water pipeline.

The Zambezi pipeline and the greening of Matabeleland North

Most of the proponents of the scheme have argued that it could irrigate substantial areas of Matabeleland North, release Bulawayo's southern dams for irrigation use in Matabeleland South, meet livestock needs and assist with mining requirements. All of these benefits would address the problem of land degradation in communal lands by relieving pressure on the dry lands and bringing increased wealth to poverty-stricken areas.

However, there is considerable scientific uncertainty over the potential irrigable area resulting from the pipeline. One study concluded that the poor soil quality in Matabeleland North offers little opportunity for irrigation (Studioemme and Geoscience 1983), while supporters of the scheme estimate that over 60,000 ha of potentially irrigable land could be developed, with direct and indirect employment creation for between 150,000 to 300,000 people and economic value to the nation of between 2 per cent and 6 per cent of GDP (Pearl 1990; Bloch 1992). The cost would be justified against the alternative of constant drought relief programmes and the long-term costs of worsening land degradation.

However, historically irrigation has competed with city requirements for water and there is disagreement even among MZWP trustees over the extent to which the pipeline might overcome such choices. Some, including Alderman Constandinos, view the pipeline as essentially a means of providing water for Bulawayo, while others, notably Dumiso Dabengwa, see the wider economic benefits as crucial to the argument for the project. Dabengwa's vision embraces not just irrigation but also mining, manufacturing and service industry in the pipeline corridor, releasing the dams in Matabeleland South for further irrigation and extending the scheme to cover the western part of Matabeleland South.

MZWP galvanises local opinion

The first public announcement of the MZWP appeared in the local newspaper, the *Chronicle*, in October 1991. The advertisement stated that only the Zambezi pipeline could save Bulawayo and meet the hopes of Matabeleland's 3 million people, and only local people could make this a reality by donating money. The money did in fact roll in. Very quickly individuals, companies and the City Council pledged considerable sums, in the council's case Z$5 million. In some factories and offices regular weekly collections were started. The *Chronicle* reflected this surge of popular support, adding to the sense of a beleaguered community helping itself.

The Department of Water Development (DWD) remained unmoved. In a statement which was to be repeated several times, one official said 'We must hope for a good [rainy] season!' The government's strategy towards Bulawayo's water crisis appeared to be based on the principle that urban areas should access those water sources nearest to them, gradually moving further afield. Thus government was looking to move to the northern watershed of Bulawayo, first to the Gwayi-Khami, then building a dam at Gwayi-Umgusa and finally developing the Gwayi-Shangani (see Figure 5.1). This last dam would cover an area exceeded only by Lake Kariba. This policy effectively represented a gradual, piecemeal development of a northern pipeline that would eventually stretch to the Gwayi-Shangani, less than 100 km from the Zambezi.

In October 1991 Queen Elizabeth II visited Bulawayo and pledged the support of the British government in finding a permanent solution to the water problem of Matabeleland. Locally this was received as an endorsement of the Zambezi pipeline project and confidence was boosted.

Impacts and developments in 1992

The end of 1991 brought the first official endorsement of MZWP's ideas. The Second Five-Year National Development Plan stated that a feasibility study had been initiated (Government of Zimbabwe 1991). This seemingly major shift in the government's stance partly reflected a change in the format of the plan, which now incorporated provincial development strategies. The Matabeleland North strategy included an uncompromising adoption of the Zambezi pipeline project. Despite this, little progress seemed to be made. As continued lack of rain in the second half of the 1991–2 season led to the realisation that a major drought was occurring, local efforts to secure support and funding for the project intensified.

A local department store erected a huge billboard with a countdown of the number of days water supply left. A locally-sponsored television programme appeared weekly for several months, entitled *Bulawayo Must Not Die*, sensitising local people to the need to conserve water and promoting the Zambezi pipeline idea. Several local companies contemplated relocating their business and some actually moved.

Early in 1992 the City Council determined a number of emergency measures designed to see the city through to the next rainy season. These included tighter water rationing, heavy penalties for over-use, an emergency programme of drilling boreholes, and finally tapping the Nyamandhlovu aquifer, some 40 km north-west. For the latter two programmes the City Council requested government assistance through loan and grant finance. The President declared Bulawayo and Nyamandhlovu a Water Shortage Area.

By October 1992 some 260 new boreholes had been drilled within the city area, with an 82 per cent success rate, the first major visible sign that the authorities were taking steps to provide some emergency water supplies should the dams run out. Utilising the Nyamandhlovu aquifer meant sinking boreholes and constructing reservoirs in that area, a task undertaken by the government, and then constructing pumps and a 47 km steel pipeline. The City Council later complained that little of the estimated costs had been released by government.

The idea of accessing water from Nyamandhlovu was seen as a desperate measure. Nyamandhlovu is a commercial farming and market gardening area providing essential produce for Bulawayo. No hydro-geological survey had been completed and it was not known how much water was available in the aquifer, whether there was a regular recharge, and hence whether the aquifer could both sustain these businesses and meet Bulawayo's emergency needs. The initial plan was for the aquifer water to supplement the city's supplies from November 1992 in case the rainy season failed. In the event, it did not come on-stream until March 1993, one year after the commencement of work. Nevertheless, the project was, in retrospect, undertaken with commendable speed and efficiency. So far as supporters of the Zambezi pipeline were concerned, the first 47 km had been put in place.

The development of the city's emergency programmes did not deflect the MZWP, nor its outspoken supporters. An increasingly critical tone against the government became apparent as it became clear that the city's survival was precariously balanced. Dabengwa threatened to resign from the government if it did not provide a lasting solution to the region's water problems.

The problem lay in a breakdown of trust. The City Council, MZWP and large sections of the Bulawayo community were convinced that the government was reacting slowly to their predicament because of an anti-Matabeleland sentiment. On the other hand, the government, certain ministers and local ZANU-PF politicians felt they were being unfairly criticised and bypassed by a populist and increasingly tribalist faction.

Throughout 1992 there was considerable worldwide interest in the drought, especially the plight of Bulawayo. MZWP produced a video publicising the situation and distributed it worldwide. It commissioned a series of pre-feasibility studies, funded by the Norwegian aid programme NORAD and the Swedish International Development Authority (SIDA). The studies were severely criticised by a subsequent Price Waterhouse assessment, but served to maintain a growing body of opinion that the pipeline project was feasible and practical (MZWP 1992/3).

Alderman Constandinos made several visits to the UK. The then Overseas Development Administration was keen to assist in the field of water conservation and undertook a study. Various finance houses were prepared to consider funding, but needed official sponsorship of the project by the Zimbabwean government. Constandinos also visited Sweden and secured a promise of Z$12 million towards a full feasibility study. The Swedes were strongly influenced by the level of community support for MZWP and the substantial funds raised locally.

From March 1992 the ruling party and the government attempted to regain some control over a situation in which neither exercised any real leadership. In July 1992, visiting Bulawayo on a high-level delegation reviewing progress on the Nyamandhlovu aquifer project, Vice-President Joshua Nkomo complained about criticism of the government. On the same occasion, Joseph Msika, Senior Minister of Local Government, Rural and Urban Development, and National Chairman of ZANU-PF, said that 'Government, in principle, accepts the offer of grants to the MZWP but certain legal requirements must first be fulfilled.' Members of the government were preparing a constitution for a new officially sanctioned body to undertake the feasibility study and generally oversee the project. This was received by some sections of the local community as a ploy to take over the MZWP.

The year 1992 ended on a more positive note as heavy rains saved Bulawayo from possible evacuation with twenty-two days supply left and the Nyamandhlovu pipeline some months away from completion. The 1992–3 rainy season saw some major inflows into Bulawayo's dams so that, by the end of March 1993, water rationing was lifted.

The slow genesis of MZWT and of the Feasibility Study, 1993–4

MZWP continued to operate during 1993, despite its possible eclipse by the proposed new body, to be called the Matabeleland Zambezi Water Trust (MZWT).[1] By 1994 the original MZWP had collected over Z$3 million in addition to the City Council pledge of Z$5 million. The government took some time to approve the constitution for MZWT, and the feasibility study was also delayed. In the meantime there was an increasing feeling in Bulawayo that the government was prevaricating and did not fully support the Zambezi pipeline idea.

In May 1993, at an international conference organised by the World Conservation Union (IUCN) on the Zambezi river system, the DWD's Director of Planning outlined a long-term strategy for harnessing water for urban authorities in Zimbabwe, but remained sceptical about the benefits of a Zambezi pipeline. Other speakers emphasised the problems of other countries seeking to take water from the Zambezi and of securing international agreement, though there was little opposition to the pipeline scheme from other countries' contributors. One of this chapter's authors, Roger Mpande, delivered a paper supporting the project on behalf of the Zimbabwe Environmental Research Organisation (ZERO).

The mounting sense of frustration was articulated by Eric Bloch, who accused successive governments of neglecting Matabeleland (Bloch 1993), and facetiously suggested, at a meeting organised by the Congress of Zimbabwean Industry (CZI) in Harare, in September 1993, that:

> Government would not be motivated to do anything to resolve the water crisis unless and until either three million Shonas are relocated to Matabeleland or, in the alternative, that as part of government's policies of decentralisation, His Excellency the President is so relocated.

Bloch's remarks attracted accusations of racism and risked a backlash from government. In October 1993 Dabengwa appealed to Bulawayo residents and others in the region to stop criticising the government over the Zambezi pipeline project, stating that further criticism might be counterproductive.

In October 1993 the DWD convened a meeting of interested parties, including MZWP and Bulawayo City Council, to report on the government's response to Bulawayo's water crisis. A department spokesman outlined progress on the design of the Gwayi-Umgusa and Gwayi-Shangani dams and hoped soon to see a start on the feasibility study into the Zambezi pipeline. Of the potential strategies to be investigated, the department's preferred option appeared to be successive development of dams on the Gwayi River system before ultimately sourcing the Zambezi.

MZWT was officially launched in July 1994. Like MZWP, it was an

NGO, but with strong involvement from the government and the ruling party (ZANU-PF). The structure of the new trust made it a representative body in a way that MZWP was not, since its general assembly comprised five members from each of several 'sponsoring organisations', namely: Bulawayo City Council, MZWP, Matabeleland Chamber of Industry, Zimbabwe National Chamber of Commerce, Matabeleland Action Group (a group of ZANU-PF politicians), ZANU-PF Bulawayo, ZANU-PF Matabeleland North, ZANU-PF Matabeleland South, Commercial Farmers Union, Zimbabwe Farmers Union and the government.

The composition of MZWT's board and its officers showed both government and ZANU-PF influence and a degree of local representativeness. The patron was Joseph Msika, a senior national party figure, while the chairman was Dumiso Dabengwa, also a senior party figure but known as a leader of MZWP. The secretary was the City Engineer of Bulawayo, while the treasurer was from CZI. The other eight board members included one from ZANU-PF and the Governor of Matabeleland South together with three from commerce and industry, two farmers' representatives, and the Mayor of Bulawayo. These were senior representatives of their respective organisations rather than self-selected activists. By the end of 1994 this board had met nine times and the general assembly twice. Some continuity was provided by the fact that MZWT employed Kingsley Dube, who had been chair of the Sandwich Club at the time of the Pierson 'fact paper' and involved in the MZWP campaign since its formation, as its coordinator.

The old MZWP remained as one of MZWT's sponsoring organisations, but had no clear role left to play. It still sought donations, but these reduced significantly after the formation of MZWT. However, for two years the old MZWP trustees refused to release funds to MZWT. An initial grant from Bulawayo City Council enabled MZWT's secretariat of two to function for the first few months, but by the end of 1994 the money had run out and funds expected from the Irish government had not arrived, leaving the organisation severely hampered.

Elections and the Feasibility Study, 1995

Within Bulawayo and Matabeleland the 1995 drought was in many respects as severe as that of 1992, providing MZWT with continued ammunition for its case to tap the river Zambezi. However, compared to 1992 there was less general hostility locally towards the government, which by now supported the aim of solving the Bulawayo water crisis through a major project. Thus it can be argued that MZWP's confrontational strategy had worked, leaving the way clear for MZWT to adopt a more collaborative approach to secure final agreements.

The lead-up to the 1995 parliamentary elections also provided a major opportunity for local protagonists of the Zambezi pipeline project to secure high level endorsement. In February, President Robert Mugabe re-affirmed the government's commitment to providing reliable water sources for Matabeleland. In April MZWT arranged for the President to launch its fundraising campaign whilst winding up a ZANU-PF election rally in Bulawayo. Msika, as Patron of MZWT, called on the government to implement the Zambezi water project and not listen to civil servants stating that it was too costly, concluding with the following statement:

> Comrade President, Harare is developing in great strides and Bulawayo is lagging behind because there is nobody who wants to invest his money in an area which is dry. Make your word a reality – that Matabeleland will not die!

Another source of local frustration was the considerable delay in starting a feasibility study. Not only must the determination of the government to subsume MZWP within the new MZWT have delayed progress, but there was then a tussle for the control of the feasibility project between the government and MZWT. In the end the government won the day and established an inter-ministerial committee, which as a compromise included representatives from MZWT.

Eventually, in April 1995, a Swedish engineering company, SWECO, linked with a local company, Hydro-Utilities, began the task of producing feasibility studies in two main stages. A preliminary study was to assess alternative pipeline and/or canal routes and come up with a preferred strategy. The second stage was to provide a feasibility report on the preferred strategy as well as an Environmental Impact Assessment (EIA).

Towards the end of 1995 MZWT announced the formation of a company under its control in which the public and financiers could invest. This step marked a public acknowledgement that the government was unlikely to fund the new project in full, and that private and donor funds would be needed.

Presidential elections and the Feasibility Study finalised

One of the best rainy seasons in Zimbabwe for at least a decade occurred at the beginning of 1996. Bulawayo's supply dams swelled to over 40 per cent full and water rationing was eventually relaxed in March 1996.

The preliminary feasibility report was presented to government at the end of 1995. It stressed the need for interim measures to safeguard and recycle existing water supplies and to consider utilising local aquifers. Second, it proposed continuing to move northwards gradually via new dams, similar to DWD's existing policies. Finally, it endorsed the idea of eventually tapping the river Zambezi from the mouth of the Deka as technically feasible and financially viable. The total cost of all its proposals was estimated as over Z$6 billion (SWECO/Hydro-Utilities 1995). The government endorsed the report and confidence increased within MZWT and the

Bulawayo community that progress could be made. It was anticipated that work on the pipeline could start by mid-1996, immediately after the submission of the final feasibility report. This was also a presidential election year. In a campaign speech in February, Mugabe said that the construction of the Gwayi-Shangani dam should be started as soon as possible.

The collaborative strategy of MZWT was demonstrated in the way it received the preliminary feasibility study and the President's amplification. A confrontational approach might have deplored the report and government's response as failing to appreciate the need for rapid action on both the Gwayi-Shangani dam and its link to the Zambezi. Instead, MZWT chose to praise the report, the Cabinet and the President, presumably to encourage progress and to retain a good relationship with the government.

The final drafts of the feasibility study, with the EIA, were presented to the government in September 1996. A notable feature was the clear rejection of the Gwayi-Umgusa dam in favour of moving directly to build the Gwayi-Shangani dam and associated pipelines and pumps (SWECO/Hydro-Utilities 1996; SWECO/PlanAfric 1996). However, government acceptance of the feasibility report could not lead directly to the start of engineering work, not least for lack of immediately available resources. Heightened efforts would need to be made to raise the necessary funding. Further studies would be required on the economic, planning and institutional issues arising from the project and funds had been secured for this purpose. The possibility of developing the project through a build, operate and transfer (BOT) arrangement with a private company or consortium were being explored.

The government's endorsement of the feasibility reports demonstrated clearly that it now recognised the importance and urgency of action towards water security for Bulawayo and Matabeleland. However, during the course of this struggle by local people to force this change, the government had changed its policy generally towards social and economic initiatives of this kind. Whilst its stance in 1991 had been that these were matters for government, in 1996 it accepted that such developments should come from the beneficiaries themselves. The government would act as a facilitator, set standards and monitor performance, but it expected civil society, especially the private sector, to come forward with initiatives and resources. In part this reflected a reality in which government now had very little money to spend on large capital projects, but it also reflected moves to greater decentralisation and local self-reliance. Thus in many respects the future success of this project now rested with MZWT and the local communities in Bulawayo and Matabeleland.

Analysis

This was a case where government policy shifted – from denying the importance of the Zambezi as a potential water resource for Bulawayo and

Matabeleland to including it as an option in its national water development planning. Over a period of about five years, from 1991 to 1996, there was also a shift from an insistence on government control and management of national water development and supply to an acceptance that other organisations could play a part in such development. To what extent were these shifts the result of NGO influence, particularly the campaign by MZWP? To the extent that the strategies employed by the MZWP and other NGOs involved were effective, what were the particular circumstances which allowed this to happen?

This case also showed clear limits to the influence of NGOs. MZWP was able to push the issue of water for Bulawayo and Matabeleland on to the national planning agenda, but not to determine the particular solution. And although non-governmental agents gained a potential role in water development, MZWP was unable to secure a role for itself.

Here we discuss six factors which we suggest explain both the influence of NGOs and the limits to that influence in this case. These correspond closely to the factors suggested in Chapter 1 as explanations for NGO policy influence generally. The first three relate to aspects of the policy, in this case water development policy in Zimbabwe, and why it might be amenable to influence. The last three are about MZWP and the other NGOs involved and the reasons such organisations might have been able to achieve influence. We do not treat the political context separately here, though when we consider the NGO strategies (the fifth factor below) we do so in terms of their appropriateness to that context.

The nature of the issue

As an environmental and resource management issue this was unusual in that it did not involve a straightforward tension between the use of a resource for development and the depletion or environmental degradation that might be caused by its use. On the contrary, it was generally agreed that the proposed use of water from the Zambezi would cause minimal disruption to the river itself while potentially ameliorating or reversing degradation and drought, and assisting both industry in Bulawayo and agricultural development throughout Matabeleland.

Effectively two questions were involved. One was how to develop water resources, particularly the river Zambezi. The other was how to combat drought and land degradation. Any tensions were about the appropriate level at which these questions should be considered – local, national or regional – and whether it was legitimate for various organisations and institutions at each of these levels to have any involvement or even control over the planning of resource use and development.

It was suggested in Chapter 1 that NGOs are least likely to have influence on questions which are of central economic and political importance to a government and the interests it represents. We can consider, first, the water resource of the river Zambezi, and, second, combating drought and land degradation in Bulawayo and Matabeleland, in terms of their political and economic importance. First, the Zambezi must be considered a national resource of strategic importance and as such the government could be expected to do all it could to maintain national control over its use, and to oppose absolutely any threat to define its use in terms of local needs. On the other hand, the problem of drought and the need for regional development in Matabeleland may not be of central importance in national economic terms. Politically, however, it is more crucial given that disaffection could lead to a regionally-based threat to the government and the ruling party's control. Thus combating drought in Matabeleland is a field where any projects and resources from local and international NGOs tend to be welcomed, since it is important that something should be done, but government itself lacks the capacity to deal with such problems directly in every locality throughout the country.

In this case, an NGO proposed to combat drought locally by developing what the government insisted was a national resource. Not surprisingly, the response was mixed. The government could not ignore the combination of local NGO pressure and international promises of resource aimed at regional development and combating drought, but national control had to be maintained over the resource (the Zambezi). We suggest this explains the government's slow response and its acceptance of the idea of tapping Zambezi water only after it had regained some control over the local organisational set-up in Bulawayo, and only as part of national water resource planning.

The government and its DWD as targets for influence

Zimbabwe inherited a bureaucratic and technocratic form of management of national resources from the colonial administration and this fitted well with the national development planning system adopted after independence. In principle in a field such as water development, national planning should foresee needs and invest to provide for them before shortfalls arise, and should do so in a way which avoids inequities between subnational geographic regions or even redistributes resources in order to reduce existing disparities.

In the case of the water crisis facing the city of Bulawayo and Matabeleland, the established institutional response was slow and in some quarters seen as reflecting a bias in the allocation of central government development funds towards other parts of the country.

The DWD is responsible for the planning and implementation of water supply projects throughout the country. Its activities include the identification, technical planning, designing and costing of new sources of water. Its guiding principles include the ultimate aim of exploiting fully the water resources of all internal rivers, the objective of building one medium-sized dam in each province each year, constructing small or medium-sized dams only where a limited local demand exists and there is no potential for a large dam, major dams to be constructed in each province where there is a proven demand and potential for major expansion based on irrigation and agro-based industries. Provision for the growth in demand in towns and cities should be made to ensure that their expansion is not inhibited, through dam construction preferably on rivers close to the urban centres, with the costs spread nationally. For new projects affecting urban settlements, the department has to work closely with the respective town or city's engineering services department.

Procedurally, new water supply projects have to go through a number of stages before implementation. After initial project identification, design and costing, by the DWD, the Cabinet Committee on Development has to support the project idea, as well as soliciting the views of the Ministry of Finance, the National Economic Planning Commission, and affected city councils or local authorities. This system is built on fairly inflexible principles and depends on technical resources and bureaucratic capacity. Each project is supposed to be judged on its merits, on the assumption that required resources will be available. However, there is no clear procedure for prioritising the allocation of limited resources. Internally the department admits that achievement of its objectives is influenced by the availability of technical manpower, equipment and finance as well as the ability of other government bodies to keep up their obligations in planning and implementation processes. Thus in effect only some of the projects which should be progressed according to the department's own principles receive proper resources.

In the circumstances it is perhaps not surprising that the response to the Bulawayo water crisis was so slow. The question of equity in the allocation of development funds is harder to deal with. There was certainly evidence of historical inequality, at least since independence. It can be argued that, in the absence of an institutional mechanism separating the political authority of the state from the bureaucracy, planning policy and implementation becomes simple political expression. Thus the state could embark on policies to meet immediate needs which are totally at the service of political ends, without the bureaucracy challenging such policies. One way of countering this possibility is by decentralisation and local participation. In Zimbabwe, although various government institutions are present in each province these merely assume the role of field administration to carry out central government directives – what has been termed 'deconcentration' (Manor 1995).

Thus there is both political space and justification for other agencies such as NGOs to work independently to develop a more sustainable water resources strategy for the region. They can attempt to provide technical, organisational and financial resources, and also meet the need for participation and accountability to ensure equity in resource allocation.

The difference in influence between different phases in the policy process

The policy process is often divided into phases. Chapter 1 distinguished three: agenda setting, decision making and implementation. One may expect a powerful decision maker to seek to keep control particularly over the central part of the process: how decisions are made. NGOs may thus find it harder to influence this phase of the process and relatively easier to have an effect in the agenda-setting and implementation phases.

In defining issues to be part of the policy agenda, full information needs to be available from as wide a range of sources as possible, so there is a clear role for outside interests. Thus providing information is a relatively straightforward way for NGOs, among others, to wield at least some influence on agenda setting. Similarly, although implementation has been called 'the crucial phase of the policy process in most developing nations' (Grindle 1980: 5), implementing policy will work best, and be seen to be in the public interest, if the decision maker does not have to force its decisions on others through its own resources alone, but can call on others to assist, particularly on organisations with some legitimacy within the populations affected by the policy (Charlton and May 1995). Here again NGOs find a role, and typically may be able to influence the way policy is carried out.

The case of the Matabeleland Zambezi Water Project illustrates and develops this argument. Thus MZWP was able to provide information about the options for accessing Zambezi water, through its own public relations and through the pre-feasibility studies which it commissioned. However, this was not sufficient to influence the agenda-setting phase. MZWP was obliged to adopt a confrontational stance in order to force the question of water from the Zambezi onto the national water development agenda, and was undoubtedly successful in doing so. Although the project is not yet into its implementation phase, it looks likely that MZWT may have a role to play when that occurs, if only to reassure the inhabitants of Bulawayo and Matabeleland that their interests are being taken fully into account. Conversely, the decision-making phase is where the government has been most concerned to maintain its own control over the process, insisting on a broad range of options being investigated in the SWECO/Hydro-Utilities feasibility study and maintaining a clear separation between the report of that feasibility study and its own decision on the way forward.

The legitimacy and capacity of MZWP and other NGOs and the relations between them

Since its leadership came from different elite groups in Bulawayo, it is not clear that MZWP should be called a grassroots organisation. However, it was certainly a successful local self-help NGO with mass support.

The nature of the public participation in this case could be characterised more as public involvement than as full participation defined in terms of social power, or 'increase[d] control over resources and regulative institutions ... on the part of groups and movements hitherto excluded from such control' (Pearse and Stiefel, 1979). Nevertheless, the fact that people within the two Matabeleland provinces were organising around the issue clearly conferred some power and control, since there was aggregation of individual strengths in the creation of an organisation which could intervene in a coherent fashion to preserve and enhance their interests.

MZWP brought together a wide spectrum of individuals and institutions to fight for a common cause in a society moulded by different interest groups. It was the first ever project in Zimbabwe where efforts to raise funds came from every sector of the community: schoolchildren, street vendors, industrialists, workers and individual foreign nationals. This fundraising was for a project that would under other circumstances have been funded by an urban authority and/or central government. A lot of international support was also raised, with several governments offering both financial and technical support, and neighbouring countries like Botswana and South Africa informally indicating interest in joint efforts to harness the waters of the Zambezi for the mutual benefit of the countries of the region.

Crucial to MZWP's legitimacy was the combination of support from the white business community, popular local political leaders and ordinary residents of Bulawayo. Other NGOs such as ZERO and IUCN provided special skills in support of the project. ZERO, with its regional network of environmental expertise, assisted MZWP in obtaining donor support for its own organisational infrastructure. ZERO's ability to articulate the environmental issues and its many connections with international environmental agencies also helped in meeting concerns about the impact of the pipeline on the Zambezi river. IUCN hosted the 1993 workshop mentioned above for discussion of regional plans to tap water from the Zambezi, which helped refute suggestions that there were international objections to MZWP's ideas.

It is interesting that the government's response to MZWP took the form of setting up another NGO rather than acting through the normal centralised departmental policy channels. Compared to MZWP, MZWT was not a political threat to government, making cooperation easier. Its legitimacy, at least in the eyes of government, came from the way its constitution incorporated key organisational stakeholders rather than giving an independent platform to activists. Joseph Msika, a senior Minister previously accused of being a major obstacle to the project, was now patron of MZWT and at the forefront in promoting the project alongside Dumiso Dabengwa, a leading activist in MZWP.

The mix of NGO strategies employed and their political appropriateness

Chapter 1 distinguished various approaches by NGOs trying to achieve policy influence. The Zambezi pipeline campaign was built on indirect pressure from successful consciousness-raising, which was at different times combined with all three modes of direct policy influence, namely confrontation, complementary activities and collaboration.

However, the members and promoters of an NGO do not necessarily sit down rationally to choose a strategy. In this case, a degree of confrontation was more or less forced on the founders of MZWP by the way it was formed out of frustration with perceived government inaction, although the overwhelming political dominance of ZANU-PF makes it very difficult for such a strategy to succeed. Nevertheless, MZWP was able to exert leverage from the potential of Matabeleland generally and Bulawayo in particular to be an alternative power centre. Although, since the 'shotgun merger' of ZANU and ZAPU in the Unity Accord of 1988, the leading Ndebele politicians have become senior members of ZANU-PF itself, the party leadership may still perceive a danger that if too many elements in the Ndebele population become disaffected there could be a real threat to the party's hegemony. Since the violence of the 'dissidents' era in the mid-1980s, there have been several outbursts of violent inter-ethnic conflict in other parts of Africa, so the potential threat of a challenge to ZANU-PF based on Ndebele identity is not to be lightly dismissed. While MZWP was always careful to avoid openly playing the 'ethnic card', it did succeed in making the need for water into a special, unifying, Ndebele interest, and this was a strong reason why it had to be taken seriously.

By independently raising funds and commissioning pre-feasibility studies, MZWP could also be seen to be undertaking 'complementary activities' parallel to those of the government and its DWD. MZWT continued with such activities. In many fields of development or welfare, NGOs carry out complementary activities with the blessing or at least acquiescence of government, since it is clear that many governments do not have the capacity for universal provision. Policy influence may occur when such activities grow beyond mere 'gap-filling' to developing new ways of providing services that have implications for existing policy and of which official account has to be taken. However, in the case we are considering here, the government did not acknowledge its lack of capacity in the field of water resource planning and development. At first, MZWP's actions were not regarded as helpfully filling gaps but almost as insubordination, threatening government's rightful monopoly control over national water resources and monopoly right to negotiate for funds with international donor agencies. In fact, in this case the 'complementary activities' were actually perceived as a kind of confrontation. Nevertheless, MZWP was demonstrating the potential for a new way of providing a service, in this case, water, in circumstances where the government's capacity was in fact too limited to maintain its monopoly position, and where MZWP was also gaining support from international interests.

Finally, with the formation of MZWT there was a shift in approach to a collaborative strategy. Like MZWP, MZWT was not entirely free to choose its strategy. Given its genesis and constitution, it could hardly be confrontational towards its creator, government. If there had been a complete lack of participation in the political culture of the country, one might have expected MZWT simply to act as a puppet of government and the ruling party and perhaps to block progress. Indeed, there was some criticism of apparent lack of movement, particularly from the original promoters of MZWP. In fact, both within ZANU-PF and in Zimbabwe at large there is political debate and airing of alternative views both privately and in the press. Thus the new NGO, MZWT, was able to use its insider position to continue to press the same case for the Zambezi pipeline that the old NGO, MZWP, had forced the government to recognise in the first place. The government, in turn, could now respond positively without appearing to be giving in to opposition.

How far the NGOs could frame the issue as an international one

MZWP's position was certainly strengthened by the international support it obtained. However, the benefit to be gained from internationalising the question of water shortage in Matabeleland was limited in that the issue was not in itself global in scope. Thus it could be viewed as an opportunity for investment in an economic development project or in terms of aid to a drought-stricken region, but as an environmental issue it was essentially localised, since it did not involve global mechanisms or have potential global impact.

Apart from the need for foreign aid, investment or technical assistance, there were only three main ways of portraying the issue as an international one, two of which were as likely to evoke opposition as support for MZWP's position. First, Victoria Falls is of unique value as a global heritage site, and any suggestion of possible environmental damage to it had to be countered at all costs. Second, the Zambezi is of great regional importance as a water resource, with potential for conflict between the various riparian states over any schemes for water extraction. Third, as discussed above, the water shortage relates to land degradation as a perennial issue for Matabeleland and this is recognised as a matter of international importance in the Convention to Combat Desertification (CCD – see Chapter 3). However, during the period of the MZWP campaign the CCD was still being negotiated and no direct link was made.

MZWP's success in raising its profile internationally was certainly an important factor in forcing the government to take notice of the Zambezi water issue. However, this was not a field in which the government could allow an independent organisation to be acting in the international arena.

The government was able to ensure the effective demise of MZWP by setting up its own NGO to take on some of the same functions, and MZWP's international support was not strong enough to counter this government strategy.

Conclusion: the changing role of NGOs in development vis-à-vis the state

It was five years from the formation of MZWP and the height of its campaign in 1991–2 to September 1996, when the final drafts of the SWECO/Hydro-Utilities feasibility study were presented to the government. This was a period during which the role of NGOs in development continued to increase throughout the world. In Zimbabwe it was a period in which the government was obliged, to some extent reluctantly, to move away from a virtual state monopoly on development programmes and to involve the private sector and NGOs in various forms of partnership.

Thus the formation of MZWT as a civil society organisation and its likely role in the development of the Zambezi pipeline project was in keeping with a general shift in how development programmes were undertaken. Was it pressure from MZWP that eventually led to such a role for an NGO (albeit not itself) being accepted by government? Or was this merely a reflection of the generally changing role of NGOs vis-à-vis the state in Zimbabwe as elsewhere? To the extent that the increased role of NGOs was at least partly due to pressure from NGOs themselves, it seems impossible to distinguish the two suggestions. However, we must conclude that MZWP and its campaign had a specific effect, even if it could be construed as part of general worldwide trend towards greater involvement of NGOs in all kinds of development.

Note

1 At the time, the new organisation was often called the Matabeleland Zambezi Water Project Trust (MZWPT). Since the original MZWP was also often referred to as a 'trust', this led to confusion between the two bodies, probably to some extent deliberate on the part of those setting up the new organisation. For the sake of clarity, we have consistently referred to the original independent NGO as 'MZWP' and the new government-promoted trust as 'MZWT'.

References

Bloch, E. (1992) 'Harnessing Zambezi Water for a better Zimbabwe', unpublished paper.

Bloch, E. (1993) 'Matabeleland: a sense of frustration', *The Financial Gazette*, 28 October.

Charlton, R. and May, R. (1995) 'NGOs, politics, projects and probity: a policy implementation perspective', *Third World Quarterly*, 16, 2: 237–55.

Dube, K. (1994) 'Bulawayo learns to live again', *Southern African Encounter*, 1.

Edwards, M. and Hulme, D. (eds) (1992) *Making a Difference: NGOs and Development in a Changing World*, London: Earthscan.

Fowler, A. (1993) 'Non-governmental Organizations as agents of democratization: an African perspective', *Journal of International Development*, 5, 3: 325–39.

Government of Zimbabwe (1991) *Second National Five Year Development Plan 1991–1995*, Harare: Government of Zimbabwe.

Grindle, M. (ed.) (1980) *Politics and Policy Implementation in the Third World*, Princeton NJ: Princeton University Press.

Manor, J. (1995) 'Democratic decentralization in Asia and Africa', *IDS Bulletin*, 26, 2: 81–8.

Mitchell, T.B. (1988) 'Water pumping costs from the Zambezi River to Central Zimbabwe', paper presented at a Water Resources seminar, Harare.

MZWP (1992/93) *Pre-Feasibility Reports on the Proposed Zambezi-Bulawayo Water Pipeline and Summary*, Bulawayo: Matabeleland Zambezi Water Project.

Operation Pipeline (1991) A supplement to the *Chronicle* newspaper, Bulawayo.

Pearl, A. (1990) 'Making the desert bloom', *African Review*, June.

Pearse, R. and Stiefel, M. (1979) *Inquiry into Participation: a Research Approach*, UNRISD/79/C/14, Geneva: UNRISD.

Riddell, R. and Robinson, M. (1995) *Non-Governmental Organizations and Rural Poverty Alleviation*, Oxford: Clarendon Press.

Stoneman,C. (1992) 'Necessary but not sufficient: political democracy and development in Southern Africa', paper for Conference on the New World Order, New England College, Arundel, mimeo.

SWECO/Hydro-Utilities (1995) *Bulawayo-Matabeleland Water Supply Feasibility Study, Pre-Feasibility Study*, Harare: SWECO/Hydro-Utilities.

SWECO/Hydro-Utilities (1996) *Bulawayo-Matabeleland Water Supply Feasibility Study, Main Report*, Harare: SWECO/Hydro-Utilities.

SWECO/PlanAfric (1996) *Bulawayo-Matabeleland Water Supply Feasibility Study, Environmental Impact Assessment Study*, Harare: Sweco/PlanAfric.

WLPU (1991) *Zambezi Pipeline Feasibility Study*, Pretoria: Watermeyer, Legge, Piesold and Uhlman.

6 The policy of fencing Botswana's communal rangelands

Onalenna Selolwane

In 1991 the Bostswana government adopted an agricultural land reform policy designed to develop the national beef industry to reflect official awareness of the relationship between environmental management and sustainable development and which was intended to encourage farmers to exploit the rangelands in a sustainable manner. A component of that policy was the fencing of communal rangelands. This chapter will examine the NGO reaction against the fencing policy. While Batswana[1] NGOs may have helped to raise the public profile of rangelands fencing in Botswana, they have not succeeded in achieving a reversal of the policy. However, it should be made clear that NGOs did not launch a concerted campaign against the issue, although they have made clear in different ways their opposition to the policy. The chapter examines the history of fencing policy in Botswana, the reasons why the 1991 rangelands fencing policy was introduced, the environmental and social consequences of the policy, and the reasons why Batswana NGOs did not exercise more influence on the policy.

Background to the 1991 land reform policy

The 1991 land reform measure was seen by government as an essential ingredient in efforts to arrest land degradation, enhance livestock disease control and encourage better land management practices to ensure the continued competitiveness of Botswana's beef exports (Botswana Government 1990). From the government's point of view, land degradation was happening because of the communal land tenure system which allowed people to exploit land without accountability. The policy makers specifically singled out the poor rural majorities who occupy communal areas as the major culprits in land degradation. It was argued that these people are driven by short-term interests to over-exploit natural resources for their immediate survival and that they do not consider the long-term impacts of such a strategy on their environment, with the result that rangelands are being overgrazed beyond the limit where they can recover sufficiently to sustain the beef industry. Based on this perception of communal tenure and poverty as the major sources of land degradation, the fencing policy aimed to change the tenure system, achieve more accountable resource use and encourage better land management practices.

It was intended that the policy would be implemented by providing incentives to farmers to help them comply with the guidelines for responsible and efficient utilisation of the range. Thorough socio-economic assessments of the various areas would be carried out to identify the requirements of the localities and the various actors who would be involved in this development strategy. The beneficiaries of the policy were identified as communities, individuals and commercial syndicates. The management practices to be encouraged include efficient stocking rates, rotational and deferred grazing, control of livestock movement, and various breeding programmes.

There was considerable opposition to the fencing policy from field officers in local government and some NGOs. Opponents argued that new scientific thinking indicates that land degradation is caused by changes in the climate and geological structure, not by overgrazing,[2] that fencing would compromise the capacity of Botswana's drought-driven ecosystem to recover rapidly, that privatisation would not arrest the tendencies of livestock owners to overstock, and that enclosure would erode access to land by smallholders and landless people, thus depriving them of a critical source of livelihood. In particular, opponents have challenged the efficacy of fencing rangelands in a drought-driven ecosystem. They contend that the rangelands have a built-in resilience which derives from their biodiversity. In drought periods, graze-sensitive plant cover is depleted while unpalatable, graze-resistant plant species remain to hold the soil and ensure that it is not further denuded and eroded. Also, such periods are not only characterised by low rainfall but by extreme variations in rainfall from place to place. Traditionally cattle have been moved from the worst affected to less badly affected localities, even over long distances, using various social 'lending' mechanisms.

That way, the argument goes, the system is able to make a swift recovery when the drought breaks. This resilience, it is claimed, is undermined by fencing and rotational grazing which restricts cattle movement and encourages a monoculture which favours palatable plant species that are graze-sensitive. Furthermore, opponents to fencing argue that it is not the land tenure system that is at fault because traditionally there had been better collective management of the land when the tenure system was communal. Rather the problem is perceived as deriving from the erosion of collective management mechanisms and their replacement by ineffective institutions which lease out the land without ensuring good land management practices. In particular, large cattle owners, among whom are the policy makers, have usually been reluctant to adopt good management practices such as sustainable stocking rates. Indeed, where sound environmental policy considerations have conflicted with the short-term profits in the cattle industry, policy makers have tended to side with the cattle interests (Yeager 1993; Selolwane 1995).

Opponents of the fencing policy have concentrated on two main arguments. First, fencing is essentially a land reform policy which expropriates land from the poor and benefits the rich who have been the major source of environmental degradation. Second, fencing will have a negative impact on the environment because post-drought recovery will be slow and uncertain, leading to a situation where degradation becomes irreversible. Opposition has failed, however, to translate into alternative policies or positive contributions on the part of environmental NGOs. Many reasons have been advanced for this failure. According to some NGOs, they were informed only at a late stage in the formulation of the policy after the government had completed its internal consultations; the process of consulting civil society was merely a public relations exercise to rubber-stamp the policy. At that stage, dissent was brushed aside as if it came from the ill informed, those who had little appreciation of the importance of cattle to Botswana's economy, or those who always put the interests of wildlife above those of human beings.[3]

The remainder of this chapter will explore other factors that help explain why NGOs had little impact on the development of the fencing policy. These include the nature, structure and agenda of the NGOs themselves, the centralised nature of political and economic power in Botswana, and the diffuse nature of the policy and related problems. It is also argued that some NGOs have historically been reluctant to react to fencing policies and programmes, and that this has informed their position on the rangelands fencing policy.

Fencing, environmental degradation and NGOs: the historical position

To appreciate why Botswana's environmental NGOs had such a limited role in the formulation of rangelands fencing policy, it is useful to examine the issue from a broader perspective, taking into account the history and political economy of fencing in Botswana. The rangelands fencing policy was preceded by a number of other fencing programmes, including veterinary cordon fences, arable fencing programmes, and a policy known as the Tribal Grazing Land Policy (TGLP), which was also essentially a land reform programme. It is important to examine the environmental issues raised in relation to these precursors of the 1991 communal rangelands fencing policy and to assess the extent to which these have informed the reactions of the NGO community. Below we examine the veterinary cordon fences and the Tribal Grazing Land Policy as key examples of the Botswana government's long-term fencing programmes.[4]

The vetinary cordon fences

The first state-initiated fencing policy in Botswana was the veterinary cordon fence constructed in 1896 by the British colonial government to contain the spread of the rinderpest epidemic south of the town of Palapye. Nearly fifty years later in 1953–4 a 330 km veterinary fence, the *Khukhe* (*kuke*) fence, was erected to control the movement of stock and separate cattle from disease-carrying wildlife species. In post-colonial times Botswana has constructed fences nationwide which, following the outbreak of the foot and mouth disease in 1977, were extended to 2,500 km in length by the mid-1980s. These fences were erected in order to break eastern Botswana into smaller disease control zones and to secure the border with Zimbabwe, both to prevent the illegal cross-border cattle traffic and to prevent contact between cattle and buffalo.

During the late 1980s two more fences were added to the network, the first to create a buffer zone between the Ghanzi district and the Ngamiland region, and the second to close the gap along the eastern boundary of the Central Kalahari Game Reserve. The National Development Plan of 1991–7 added two more veterinary cordon fences to the network. These fences are primarily intended to give the nation greater control over foot and mouth outbreaks and other diseases which may threaten Botswana's position in the international beef market. Until the late 1980s these fences were put up by the Ministry of Agriculture without much consultation with other ministries, such as the Ministry of Commerce and Industry which houses the Department of Wildlife and National Parks. This led to a disregard for the migratory needs of wildlife.

The combination of fencing and drought conditions led to a dramatic loss of migratory wildlife whose survival depends on access to ecosystems in different geophysical locations during the dry and wet seasons. The most drastic reduction in wildlife has been in the area that includes the Kalahari (Kgalagadi), Ghanzi (Gantsi) and Kweneng districts in the south-western part of Botswana. Here the wildebeest population has declined by up to 94 per cent since the late 1970s. The hartebeest population dropped by 83 per cent over the same period. The migratory routes of these species have been closed off completely, restricting the animals' access to alternative resources available seasonally elsewhere. Animals like the eland have suffered relatively less as the fences did not interfere to the same extent with their migratory patterns. Nonetheless, they have been reduced to half their 1970's numbers. Significantly, this south-western region of Botswana was the main area targeted for the 1975 Tribal Grazing Land Policy.

In the less arid north-eastern region which is home to species like the elephant, buffalo and zebra, the impact of fencing on wildlife has been less severe because the migratory species of animals found here have better access to permanent surface water. Although zebra and buffalo populations have declined by 80 per cent and 30 per cent respectively since the mid-1980s, the elephant population has increased, almost doubling in the past decade alone. This has raised concern because of the ability of elephants to alter their habitat. In some parts of the region the elephants have pushed the ecosystem close to breaking point in terms of its resilience and capacity for reversible transformation; the abundance and diversity of flora and fauna decrease towards the watering points used by elephants. Both in the hard hit south-western region and the relatively less degraded north-east there is evidence to suggest that, unless drastic measures are taken to create openings for wildlife migration through the veterinary fences that have been erected to protect the beef industry, Botswana's wildlife will soon be reduced to a few animals in protected areas. The official perception is that drought conditions are the major source of wildlife depletion (Botswana Government 1991a). Others have argued that, although drought has been a major factor, the veterinary fences have exacerbated the problem by making it impossible for animals to access spatially dispersed resources which are critical for their survival.

Surprisingly few environmental NGOs have reacted against the fencing programme so far. The largest and most influential NGO in Botswana, the Kalahari Conservation Society (KCS), has indicated that it is aware of the devastation to wildlife that can be caused by veterinary cordon fences and that it has opposed them in the past.[5] KCS also recognises the positive side of such fences. For instance, they accept that the buffalo fence to the west and south of the Okavango Delta helps to protect the delta from livestock encroachment. KCS, according to its chairman, would like to see this fence extended further north.

Critics of KCS argue that the organisation is incapable of offering objective assessments of the environmental impact of government policies as it has no credible independence from government and that it is in effect an extension of government into civil society. Although KCS has sponsored research on those aspects of environmental policy development falling within its range of interest, principally wildlife and the Okavango Delta (Chapter 7), it has not sponsored research on either the impact of cordon fencing or alternative methods of disease control that might allow wildlife migration, an issue that is central to KCS's mandate. At a symposium the KCS organised in 1992 to review the status of the environment in Botswana, the only paper delivered on wildlife management and conservation was on the Namibian experience with wildlife. Other NGOs concerned with wildlife include the Association of Wildlife Clubs, the Botswana Bird Club, the Chobe Wildlife Trust, the Khama Rhino Sanctuary Trust, and the Mokolodi Wildlife Foundation. Like the KCS, these NGOs do not appear to have a policy position regarding cordon fences and their environmental impact. On the contrary, it is the Ministry of Commerce and Industry's Department of Wildlife and Parks which has not only expressed serious concern over the impact of cordon fences on wildlife, but has also proposed alternative strategies to ameliorate the situation by, for example, providing wildlife corridors and animal watering points.

The Tribal Grazing Lands Policy

While the network of nationwide veterinary cordon fences has been established to control inter-regional cattle movement and cattle diseases, other fencing systems have been introduced by the Ministry of Agriculture to improve the management capacity of cattle farmers. The first of these was part of a land reform strategy, the Tribal Grazing Lands Policy, which was adopted in 1975 with the objective of privatising communal rangelands and delineating the area of responsibility of the individual farmers so that they could then strive towards improved commercial production of beef.

The policy was based on the assumption that the communal land tenure system practised by local pastoralists was a major factor in poor, low output management practices, and that communal lands could not properly support the development of the cattle industry and the beef export market. Privatisation was thus seen as a means of enabling some local people to engage commercially in the cattle industry by adopting 'improved' farming practices such as efficient stocking rates, rotational and deferred grazing, use of modern technology, and breeding programmes which relied on stricter control of livestock movement (Botswana Government 1991b). Critics of the TGLP have focused not so much on the fencing itself and its environmental impact as on the social and economic impacts of the programme, such as its endowment of exclusionary rights of land use and the opening up of territory previously closed to cattle.

Components of the TGLP such as the sinking of boreholes have not only ceded communal territory to a minority of local people, but have led to a considerable transformation of the natural resource base, depleting the diversity of plant and animal species as well as preventing the access of other users to remaining species which are still an important source of livelihood. But it is the control of grazing within the enclosures that has been seriously challenged as a resource management tool. The suitability of rangeland fencing as a strategy for enhancing resource management and improving the productivity of livestock agriculture has been questioned. As already mentioned, critics argue that enclosure can erode the resilience of the rangeland ecosystem to severe drought.

More importantly, after two decades of the TGLP experience, there is little evidence that the beneficiaries of land reform have actually adopted more efficient resource management practices. On the contrary, the evidence suggests that the majority of these farmers have considerably increased their livestock holdings beyond the estimated carrying capacity of the land, causing serious land degradation. In times of drought-induced reductions in their grazing areas, ranchers have simply transferred their animals onto common land, thus exploiting their 'dual rights'[6] in order to enable their private lands to recover. The result is serious range degradation on the commons.

Although the official point of view is that the TGLP experience has been a success that has demonstrated the capacity of Botswana to practice commercial farming, other observers have argued that the TGLP has been a resounding failure. Perkins (1991: 190) observes that TGLP has made an insignificant contribution to the productivity of large farmers who perform no better than subsistence producers on communal lands if their total output rather than just their beef output is taken into account. Another criticism is that the traditional system produces a greater number of useful products of higher value and at a lower price than the commercial ranch. Evidence suggests that the overwhelming majority of the ranchers have actually not adopted the improved management practices recommended by the Ministry of Agriculture (Selolwane 1995). The ranchers' ability to make profit may have been helped by generous subsidies from the government and the European Union, allowing them to minimise costs and increase their herd size without incurring expenditure on land management.

Policy development on land use and environmental management has thus not forced any greater sense of responsibility in these farmers' agricultural practices. Policies have not so far emphasised accountability, monitoring and impact assessment. Gradually the environmental and social costs of these management practices have become more apparent, to the point where they can no longer be ignored because of the threat they pose to Botswana's capacity to sustain its agricultural development. However, this has made agricultural policy makers even more convinced that the way forward lies in extending exclusionary land tenure systems. The 1991 fencing policy should therefore be seen as an extension of the TGLP strategy to all communal areas in the country.

Few environmental NGOs have specifically reacted to the TGLP in the two decades of its existence. Some critics have alleged that, because of its primary concern with wildlife, KCS has not recognised the fencing of the rangelands as an environmental issue at all since it does not pose any direct threat to wildlife. However KCS has refuted this.[7] While KCS acknowledges that fencing can lead to overstocking and land degradation, it is not an issue on the organisation's agenda. This illustrates the sectoral rather than holistic approach common to many Batswana NGOs. In many cases, Batswana NGOs do not recognise a clear relationship between their defined area of concern and broader policy issues, hence they do not perceive a problem that requires their attention. Few of the smaller environmental NGOs have developed a policy stand towards rangelands fencing, despite the mounting evidence of environmental impact and social consequences. Criticisms of the fencing policy from the NGO community are usually individual reactions rather than organisational positions.

The 1991 fencing policy

The 1991 fencing policy, which forms part of the 1991 National Agricultural Policy, was not therefore the first such policy with environmental impacts of which Batswana NGOs were presumably aware. Yet NGOs have made little attempt to influence the policy. Part of the reason for this concerns their organisational structure. Most of these NGOs are small, localised institutions which were created in the 1990s. They have limited resources (human and financial), and therefore tend to have a narrowly-focused environmental agenda. Even the bigger NGOs like the Kalahari Conservation Society tend to have a narrow agenda, reacting to environmental issues only when they directly impinge on their interests, such as wild animals, forestry and veld products.

During the course of this study the principal environmental NGOs that indicated in interviews more than a passing awareness of the 1991 agricultural policy and its fencing strategy included the Forestry Association of Botswana (FAB), the Forum for Sustainable Agriculture (FONSAG), Thusano Lefatsheng and the Botswana Society. These NGOs perceive the fencing policy as an unpopular programme which has been forced through essentially because it is a government initiative. FONSAG believes that their opposition to the programme may have been a factor that delayed its implementation while FAB think they have probably forced the government to re-examine some of the issues more carefully and carry out further research. Nonetheless. the policy is being implemented, with areas identified, assessed for environmental impacts and designated for individual, syndicate and communal ranches, although there have been some delays due to time-consuming land use assessments and mapping.

Many NGOs attended the second National Consultative Conference (24–8 June 1996) to review progress and finalise the implementation of the fencing policy. Amongst the issues reviewed at this meeting were dual rights, economic and financial issues and the extent to which financial institutions might contribute to the implementation of the policy. However, the NGOs participated in this exercise primarily to react to proposals, rather than to contribute any of their own. In the interim period between formulation and implementation, NGOs confined themselves to declaring general reservations. In interviews conducted at the time of the second National Consultative Conference, many environmental NGOs appeared uncertain of the exact status of the policy, illustrating their peripheral interest in rangelands fencing. While they believe they share a common position of opposition to the fencing of the rangelands, NGOs have not demonstrated a strong, united stance against the policy.

The Forestry Association of Botswana claims that its main contribution to the fencing debate has been pushing for a government review of the environmental impacts of the precursor TGLP.[8] This proposal was made at the first National Consultative Conference in 1990 after which little effort appears to have been made to pursue the issue. Around this time the NGO Thusano Lefatsheng was involved in regular contact with the government through attendance at committee meetings. However, Thusano Lefatsheng as an organisation has no clearly articulated policy stand on fencing. Like FAB, Thusano Lefatsheng was strongly represented at the 1990 National Consultative Conference on agricultural fencing where it voiced objections to the fencing package while emphasising that the policy is principally an agricultural policy with no direct bearing on the interest of Thusano Lefatsheng. An official from the organisation suggested that Thusano Lefatsheng was opposed to the policy on three counts: it would be likely to concentrate land in the hands of the rich; it would deny access to veld resources for non-leaseholders; and it would lead to land degradation due to the exercise of dual grazing rights by leaseholders.[9]

Apart from expressing general objections to the fencing policy, however, none of the NGOs have offered alternative policy suggestions. Yet both the proponents and opponents recognise that already many large cattle owners have seriously affected the rights of access to natural resources of the poorer farmers and communities. Even before the new fencing policy the poor had been excluded by large farmers from certain rangelands as a result of the TGLP, making it difficult for them to procure fuelwood, raw materials and grazing space. The fact that fencing is now an entrenched tool in Botswana helps explain why NGOs have been reluctant to orchestrate a campaign against the 1991 rangelands fencing programme. As one NGO official observed, when the government bureaucratic juggernaut has been set in motion, it takes considerable pressure to force it to halt or change direction. However, when a new policy is announced, Batswana NGOs seem more willing to work together in a joint campaign. An example here is the case of the Southern Okavango Integrated Water Development Project (SOIWDP) which was halted in 1991 on the eve of its implementation (Chapter 7 of this volume). Certainly there has not been the level of pressure on fencing that was applied in the case of the SOIWDP. In other areas too the NGO community has made significant inputs in the development of environmental policy in Botswana. Indeed the very existence of environmental policy in Botswana in terms of the National Conservation Strategy – as well as specific policies on forestry, veld products, wildlife resources and the inclusion of environmental considerations in development policy and planning – is a direct consequence of NGO initiatives. The government itself acknowledges that environmental policy development has been largely influenced by NGOs.

The limited political base of environmental NGOs

NGO interventions in policy developments are directed at both the policy makers (mainly government) and members of the public whose activities have a cumulative impact on the environment. In many respects NGOs confine themselves to damage limitation through informed lobbying, the development of schools' curriculum and practical work aimed at developing and facilitating conservation projects. In so far as NGOs seek to intervene in environmental and development policy and practice, their efforts are usually directed at government. It has become a norm of political life in Botswana that, until the government has formulated policy on an issue, it is virtually impossible for anyone else to act on it with any degree of effectiveness. It is generally assumed that the government must first adopt a national policy and then direct other actors on how it should be conducted. Development projects are not considered if they do not have government's blessing. Thus, despite Botswana's liberal democratic culture, civil society groups are weak both as political actors and development managers. It is also rare for Batswana NGOs to lobby the private sector to take action to improve the environment.

A major weakness of Batswana NGOs is that they are not deeply-rooted in local communities. They tend to be patronised by educated elites and academics, often from the expatriate community. This weakens their capacity to involve ordinary people in decision-making and environmental management, resulting in the exclusion of local knowledge and concerns in the formulation of environmental agendas and strategies. In the past, the preponderance of white expatriate members has exposed some environmental NGOs to charges of foreign interference in Botswana's internal affairs, especially when development policies have been criticised for causing environmental damage. In particular, since many of these NGOs are concerned with wildlife rather than the broad spectrum of environmental issues, they are also vulnerable to charges of putting the welfare of animals before those of the people of the country.

Over the years the NGOs have striven to attract Batswana into their membership. A few are now headed by black people while others, such as the Kalahari Conservation Society and the Botswana Society, are patronised by prominent politicians, society people and corporate members. As a result the research activities of these NGOs have moved more towards national development and the role of local communities in resource management. However, the government remains cautious of the NGO sector, and there is a general perception in the state bureaucracy that these NGOs may create alternative power centres that could lead to divisions in the nation, thus undermining nation building.

The same lack of strength can be seen in the political party system, where the opposition has been very weak relative to the ruling party. Yet this weakness cannot be attributed to political persecution or the threat of violence as has been the case in many other African countries. So while not discounting government hostility as a factor in the weak development of organisational life, it is also important to assess the weak status of NGOs within the broader context of a general lack of power diversity in the country, despite the presence of factors that ought to nurture a robust political and civic culture. Observers have usually attributed the weakness of civil society to two main factors: the high politico-bureaucratic centralisation of government; and the favourable economic circumstances which have enabled the government to deliver basic development services to most communities. Botswana has experienced a high and sustained rate of economic growth over its post-independence history that has been unmatched elsewhere in Africa. This economic growth has put considerable resources at the government's disposal, and these have been used wisely to provide basic social services and infrastructure to most rural communities.

Economic prosperity has also consolidated the post-independent state bureaucracy, and one consequence of this is that the government has undermined traditional authority structures such as the *kgotla* and eroded the decision-making power of local communities, including community management and control of natural resources. Local control has been replaced by centralised control, exclusionary use of natural resources, and an emphasis on private and individual interests. The spirit of community has been eroded in spite of the proliferation of state-instituted mechanisms, such as Village Development Committees, to provide a meeting ground for government institutions and the communities. Participation of local communities in these committees is based on their position as junior partners in decision-making relative to officers of the state. The development of new forms of organisational life, such as that represented by NGOs, has followed basically the same path of subordination and marginalisation. NGOs with a large expatriate membership lack local support, and with neither a grassroots foundation nor the economic base to purchase political space, such NGOs remain weak, with a narrow constituency and limited political influence.

However, NGOs may achieve some influence where their campaigns do not directly challenge the economic gains of the 'cattle barons', many of whom have close ties to the political elite. While the Batswana elites have diversified away from cattle into commerce and the bureaucrats are no longer primarily cattle barons (Molutsi 1988), the priorities of Botswana's policy makers continue to be heavily influenced by their interest in safeguarding the lucrative cattle industry, from which they derive a substantial part of their revenue. There is an intersection of power between cattle ownership on the economic side and the occupation of government office on the political side, suggesting a coherent power base which makes it difficult for those outside this core to challenge development decisions. Cushioned by support from politicians and generous EU subsidies, the cattle barons have little incentive to look beyond immediate financial gains to wider, longer-term environmental costs and the impoverishment of those with limited access to cattle and wealth (Selolwane 1995; Perkins 1991). The result is that NGOs have limited political space to develop alternative power bases where they could influence environmental policy and development programmes. This seriously limits the scope for more effective NGO participation, for the development of a diversity of power centres, and for fully democratic governance. The more radical NGOs thus often turn to external sources of support to put pressure on the government to change. Some NGOs see themselves as the saviours and protectors of wildlife heritage in a political system where government is dominated by greedy cattle barons with such a monopoly of power that they can only be opposed by strong external forces. This perspective results in a confrontational stance towards government, and is sometimes interpreted by government as a threat to national cohesion (see Chapter 7 below). The capacity of such NGOs to influence policies in the long term can therefore be seriously impaired if they are seen to challenge the legitimacy of the state.

The nature of fencing as an environmental problem

The failure of environmental NGOs to make a significant impact on fencing policy can also be attributed partly to the complexity of the issue. For example, while it is easy to refute the arguments of some policy makers that land degradation is a problem caused by the poor (by pointing to the skewed distribution of cattle and land ownership), it is not so easy to prove categorically that fencing rangelands will exacerbate land degradation. The government claims that the fencing policy is not just directed at the poor but at all cattle owners. It is claimed that it will benefit all those who practise communal grazing and not just the commercial syndicates and individual commercial interests with a stake in the cattle industry. So when concerns are raised about the possible expropriation of the rights of access to land of certain social groups,[10] the policy makers refer to the communal rangelands component of the policy which, it is claimed, will safeguard the interests of all those who still need to use the commons.

When questions arise concerning the possibility of further encroachment on the resources of those communities which have already experienced a land squeeze as a result of the borehole syndicates, the government's response is simply that these big owners will be encouraged to move to more open space through incentive schemes.[11] Thus because the threat of dispossession is not always obvious and immediate, policy makers can deflect concerns and put their detractors off balance by pointing out supposed safeguards.

There is also considerable uncertainty over the exact social, economic and environmental impacts of fencing, despite the robust debate that has surrounded the formulation of this policy. For instance, from the policy point of view the negative impacts of the precursor of this policy (the TGLP) are usually attributed to poor implementation and lack of adequate monitoring, rather than to the policy itself. The TGLP's failures, if they are admitted, are often attributed to the mass transfer of the civil servants who should have implemented the policy to the National Development Bank which provided the loans to the beneficiaries.[12] The overstocking of cattle by large farmers is often attributed to ineffective monitoring by the Land Boards, and not to any failure on behalf of the large farmers.

The effects of fencing policy depend on many factors, such as the exercise of dual rights by the more affluent farmers, effective protection of the small farmers and those without livestock, the effectiveness of incentive packages and environmental education. The effects also depend on the monitoring and statutory powers conferred on the leasers of the land and how willing they are to exercise those powers against farmers who may be their superiors, such as senior civil servants and cabinet ministers. This makes it difficult to isolate the social impact of any particular aspect of the policy. Similar problems arise with regards to assessing the policy's environmental impact. It is not just the physical characteristics of fencing that have an impact, but also the behaviour patterns of the farmers. Will they, for instance, adopt environmentally friendly farming practices such as ensuring that biodiversity is not seriously affected, or are such compromises completely impractical under conditions of limited stocking? It is not easy to state categorically either that the policy will lead to environmental degradation and impoverishment, or that it will enhance conservation and the enrichment of all potential beneficiaries.

With such a diversity of issues to be considered and assimilated by NGOs, it is perhaps no surprise that NGO opposition to the 1991 fencing policy was muted. The fencing issue is not a clear single issue problem where the energies and strategies of those challenging it can be focused in one particular direction, as was the case with the proposed dredging in the Okavango Delta (Chapter 7).

Concluding remarks

In summary, the failure of NGOs to influence policy change in the fencing component of Botswana's 1991 agricultural strategy can be attributed to four main factors. First, NGOs have been narrowly focused in terms both of their environmental agendas and their geographical and social spread, while the policy they sought to challenge is national, encompassing a whole spectrum of social and political constituencies. Second, the central importance of cattle to the economy, particularly to the incomes of those in government power, tends to bias decision making in favour of large cattle owners and their interests. With the diminishing availability of new lands, there has been a tendency for this powerful lobby group to expropriate more land from the communal areas to support the beef industry. Third, the generalised nature of the fencing policy, and its embeddedness in a complex socio-economic network, makes it a difficult and diffuse target to challenge. Fourth, fencing encompasses a range of regions. In order for NGOs to identify the impacts of the policy, separate environmental impact assessments would be needed for all regions, an option which is far too costly for NGOs with limited resources. The state, however, has been able to call on its vast human and economic resources to analyse the issues more concretely and comprehensibly than the NGOs.

The fencing debate has left many outstanding issues. The question of how to balance commercial interests with environmental considerations is one such example. A related issue is the question of how to limit environmental degradation and increase the benefits to all stakeholders. While the fencing policy purports to benefit everyone, it is far from clear that the strategy for implementing the policy will either promote sound environmental management or ensure the equitable distribution of benefits. Furthermore, given the role of Botswana's variable rainfall in the distribution of range resources, it is not clear what measures will be taken to avoid the land degradation that occurs when cattle movement is restricted.

Notes

1 We use the term 'Batswana' to describe NGOs from Botswana. This does not imply that they are specifically devoted to the interests of the Batswana ethnic group, the great majority in the country. Indeed several NGOs represent other interests, such as the Basarwa, and others have expatriate membership.

2 Interview, Richard White, member of the Kalahari Conservation Society Technical Committee.

3 Interviews with officials in the Ministry of Agriculture and environmental NGOs.

4 Arable fences have perhaps not been as significant sources of environmental deterioration as the others, but their contribution is still worth noting. These are the fences sponsored for arable agriculture under programmes like the Accelerated Rainfed Arable Agriculture (1985), Arable Lands Development Programme (1980) and the drought relief programmes. There is evidence to suggest that, because the programmes encouraged farming and fencing during drought years when the vegetation was already suffering from reductions, they encouraged overuse of trees and shrubs since the farmers received grants for fencing. Together with population pressure, the expansion of human settlements and the increasing tendency for rural people to reside permanently on their lands, this has led to increasing encroachment into wildlife territory as well as the depletion of natural resources such as trees, medicinal plants and wild foods.

5 Interview with the Chairperson of KCS, Louis Nchindo.

6 'Dual (grazing) rights' occur when a person owning private grazing land also retains a customary right to access common grazing land.

7 Seminar comment, Keith Leggett of KCS, GECOU/COMMUTECH, seminar on *Environmental Policies and NGO Influence: Land Degradation and Resource Management*, Harare, 26–8 June 1996.

8 Interview with the Director of FAB, Kesileng Walker.

9 Interviews with Thusano Lefatsheng officials (conducted separately by Alan Thomas and Onalenna Selolwane).

10 The Botswana Christian Council, for instance, and the First People of the Kalahari are two NGOs which have been particularly concerned for the welfare of Basarwa hunter/gatherer communities whose land rights were seriously eroded in the first grazing lands policy.

11 The evidence of the first TGLP is that most of these big ranchers never moved from the communal areas to the designated ranches, but have simply benefited from the largesse of the state by effectively controlling the commons by virtue of their borehole ownership and private fencing.

12 Interview with the Chief Agricultural Economist, Mr Howard Sigwele.

References and sources

Botswana Government (1990) *National Policy on Natural Resources Conservation and Development*, Gaborone: Government Printer.

Botswana Government (1991a) *Government Paper No. 1 of 1991: National Policy on Agricultural Development*, Gaborone: Government Printer.

Botswana Government (1991b) *National Development Plan 1991–97*, Gaborone: Government Printer.

Devitt, P. (1978) 'Drought and poverty', in M.T. Hinchey (ed.) *Proceedings of the Symposium on Drought in Botswana*, Gaborone: Botswana Society.

Good, K. (1993) 'At the ends of the ladder: radical inequalities in Botswana', *Modern African Studies*, 31, 2: 203-230.

Hichcock, R. and Holm, J. (1993) 'Bureaucratic domination of hunter gather societies: a study of the San in Botswana', *Development and Change*, 24: 325–38.

Holm, J. and Molutsi, P. (eds) (1989) *Democracy in Botswana*, Gaborone: Macmillan Botswana.

McGowan International and Coopers & Lybrand (1987) *National Land Management and Livestock Project: Incentives/Disincentives Study Vol. III*, Gaborone: McGowan International and Coopers & Lybrand.

Molutsi, P. (1988) 'The state, environment and peasant consciousness in Botswana', *Review of African Political Economy*, 42: 40–7

Molutsi, P. and Holm, J. (1990) 'Developing democracy when civil society is weak', *African Affairs*, 89, 356: 323–40.

Perkins, J.S. (1991) 'Drought, cattle-keeping, and range degradation in the Kalahari, Botswana', in J.C. Stone (ed) *Pastoral Economies in Africa and Long Term Responses to Drought*, Aberdeen: Aberdeen University African Studies Group.

Selolwane, O. (1995) 'Fencing Botswana's commonage: a recipe for environmental disaster or conservation?', *PULA*, 9, 2: 90-117.

Yeager, R. (1993) 'Governance and environment in Botswana. The ecological price of stability', in S.J. Stedman (ed.) *Botswana: The Political Economy of Democratic Development*, Boulder: Rieriver.

wa-Githinji, M. and Perrings, C. (1993) 'Social and ecological sustainability in the use of biotic resources in Sub-Saharan Africa', *Ambio*, 22: 2–3.

7 The campaign against the Southern Okavango Integrated Water Development Project

Alan Thomas, Onalenna Selolwane and David Humphreys

This chapter analyses the NGO campaign against the proposed Southern Okavango Integrated Water Development Project (SOIWDP) in Botswana. The SOIWDP was a planned large-scale engineering project involving the dredging of one of the main waterways of the Okavango Delta, sometimes described as the world's greatest oasis. The swamp, water channels and hundreds of islands form habitats for a variety of birds and animals, so that the Okavango Delta is world famous for its wildlife. It also provides livelihoods for over 100,000 people of a variety of ethnic groups, including Yei, Batawana, Herero and Europeans.

Plans for engineering projects to utilise the water of the Okavango pre-date independence. One South African MP in the post-war Smuts government argued that, if the then Bechuanaland were incorporated into South Africa, 'we could then develop the mighty Okavango River and, as it were, create a new province ... out of the desert-like Kalahari' (Hyman 1974: 73). What became the SOIWDP was selected as the least environmentally damaging of several recommended schemes in the 1970s. By 1985 terms of reference had been approved for a feasibility study, undertaken by the Snowy Mountains Engineering Corporation (SMEC) of Australia and co-financed by the Botswana and Australian governments. The main emphasis was on 'optimum use of land and water resources culminating in increased agricultural production benefiting the region and the nation' (SMEC 1987a: ix). The SOIWDP was envisaged as a multisectoral project for meeting a wide range of goals and was planned to provide water for commercial irrigation, village-based flood recession agriculture, livestock, domestic use for the town of Maun and surrounding areas, and the Debswana diamond mine at Orapa. In fact the proposal to supply the diamond mine was dropped quite early in the political debate about the project although the connection provided ammunition for some of the scheme's opponents.

The proposed project planned the dredging and bunding of 42 km of the lower Boro River in order to increase water inflow into a 100 km long Maun reservoir, which would be impounded behind dams. A control structure at Matlapaneng would hold another reservoir. Finally, a dam

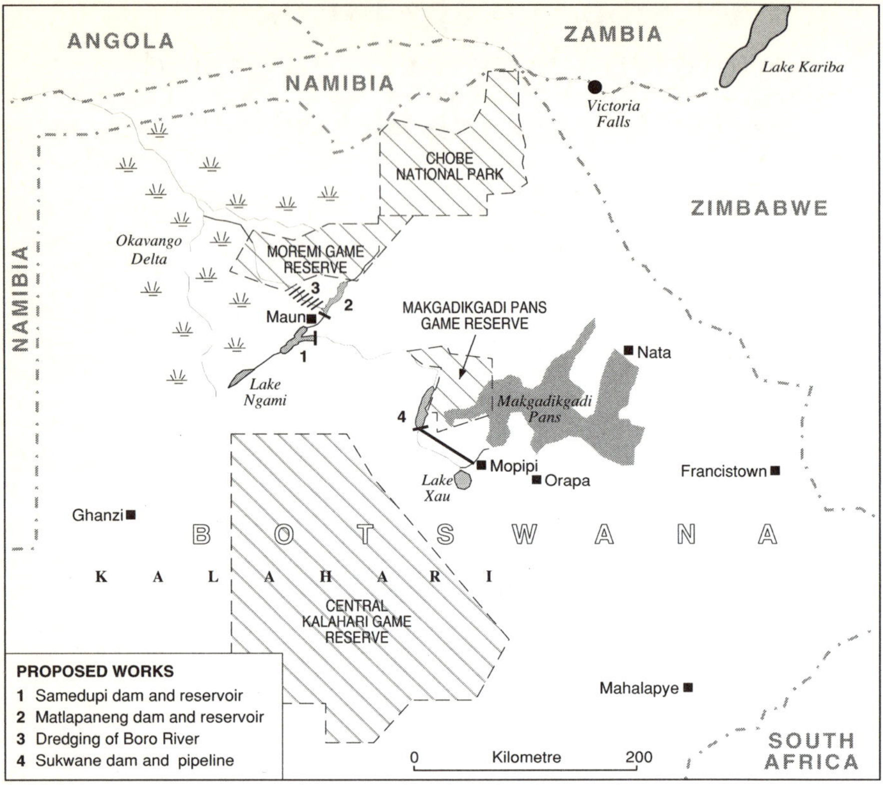

Figure 7.1 North-west Botswana showing the Okavango Delta and proposed engineering works for the SOIWDP.

would be constructed on the Boteti river, with pipelines supplying water from another reservoir to the Orapa diamond mine (see Figure 7.1). The aims of the project changed over the years it was developed. In 1988 food production was dropped as the primary aim. SMEC then reviewed the options giving priority to supplying water for Orapa and Maun (SMEC 1988). High cost estimates for some of the project components led to the Matlapaneng control structure and reservoir being taken out of the revised project design. The cabinet approved implementation of the first stage in late 1988. Local people only realised the project was agreed when earth-moving equipment arrived in the town of Maun towards the end of 1990. The campaign began when local environmentalists and others opposed to the project mobilised local communities and formed an action group which later became the Tshomorelo Okavango Conservation Trust (TOCT). International NGOs also became involved in different ways, notably Greenpeace and IUCN.

The Botswana government eventually halted work on the project in 1991, although it reserves the right to develop the water resources of the Okavango in the future. There were two interacting dimensions to this policy change. The first concerns the role of expert knowledge. The second concerns the role of NGOs. These are now examined in turn.

The role of expert knowledge

Most of the important environmental issues in Botswana, a country with low rainfall and entirely arid or semi-arid, concern land degradation and sustainable resource management. Land degradation in Botswana has been described as

> the inevitable result of the misuse of the land and vegetation resources by overgrazing and the adoption of inappropriate agricultural practices. It consists of a number of interlinked processes, including a reduction in soil fertility, erosion of the soil, depletion of the vegetation biomass and ... [the] end product is desertification.
>
> (Moyo et al. 1993: 40–1)

There is considerable scientific and political debate about these issues. It is important to understand their construction in terms of economic and development benefits on the one hand and their potential environmental impacts and risks on the other.

Until the 1970s there was no question of cattle in the Okavango because the area was infested with tsetse fly, although 'for years the cattle industry has looked with envy at the lush swamps and flood-plains of the Okavango Delta' (Lomba 1992: 7). Since then eradication measures financed by the UN Food and Agriculture Organization and other donors have allowed new areas of cattle range to be opened in the region, notably to the north-east of the Okavango Delta, now divided from it by the Northern Buffalo Fence.

Disputes about large-scale water engineering projects in arid regions such as the SOIWDP relate to issues of land degradation and sustainable resource management in two main ways. First, water is a basic natural resource in short supply in areas like the Kalahari, hence there are always competing demands and alternative proposals on water utilisation. In particular there are conflicts between proposals to use water resources for national economic development and proposals for local community-based forms of development that may be easier to reconcile with conservationist ideas such as protecting endangered species and biodiversity. Second, land degradation could be hastened or reversed depending on how water resources are managed. Both the proponents and the opponents of the SOIWDP cited the avoidance of land degradation as one of their objectives. Proponents argued that evapo-transpiration rates are so high in the

Okavango that careful water management would enable water to be used for irrigation, domestic purposes and mining without disturbing the delta ecosystem. They pointed out that if part of the water was stored in reservoirs with greater depth evaporation would be reduced overall (Porter 1989: 6). Opponents argued that water would be removed more quickly from the delta so that the Boro floodplain would dry out and become degraded, that the entire Northern Wildlife Area, of which the delta is only part, should be considered holistically (Williamson 1994: 12), that previous large-scale irrigated development projects had failed, and that the SOIWDP could have a negative impact on areas to the south of the delta proper if it encouraged over-production of cattle or agricultural products relative to the carrying capacity of the land. They argued that the water needs for irrigation and other local purposes would be better met through a combination of smaller, less intrusive measures.

An environmental impact assessment (EIA) was prepared by the Snowy Mountains Engineering Corporation (SMEC) at the same time as their feasibility study (SMEC 1987b). This predicted that the main negative impact would occur on the Boro River and its floodplain. The EIA estimated that the proposed dredging and bunding would dry out about 70 square km of floodplain with negative consequences on animal and bird wildlife including the endangered wattled crane for which the Okavango, particularly the sedgeland area of the lower Boro, provides one of the few remaining breeding areas. The dredged section would be likely to become virtually a sterile habitat so that the wilderness value of one of the main tourist boat routes into the delta would be much reduced. Considerable negative impacts on fisheries important for local communities were predicted. However, the impact would be limited to less than 1 per cent of the delta, and this led it to be judged acceptable. The anticipated economic gains were regarded as outweighing the relatively small environmental costs.

However, in a government-sponsored review of the environmental impact studies Skinner suggested that the area of floodplain likely to be affected was 275 square km and that the project might 'act as a gate opener to future incursions into the delta' (Skinner 1989: 12, 19). Subsistence economic activities carried out on the lower Boro could be displaced because of the negative effects of the project on carrying capacity, so that the total area affected could be as much as 3.5 per cent of the delta. Furthermore, demand for water might not be satisfied by the project so that there would be pressure to continue the river 'improvements' into the heart of the delta.

The efficacy of cost–benefit analysis presupposes a high degree of accuracy in the estimates of both the positive and negative impacts of a proposed project. Increasingly, however, scientific evidence in cases like that of the EIA of the SOIWDP is being questioned and it is suggested that such proposals should be analysed in terms of risk and uncertainty. Many critics of the SOIWDP have claimed that both the benefits and costs are marked by extreme uncertainty; the damage to the integrity of what is a unique ecosystem is essentially unknowable in advance, and it cannot be known for certain that the project will succeed in delivering extra water supplies. There are two kinds of argument here.

First, local people point out that the Okavango consists of shallow water and swamp spread over a wide area on top of Kalahari sands. They argue that river water would seep into the sands were it not for the hard river beds. Dredging the Boro would break up this hard surface, causing seepage and siltation, either diverting the river or requiring regular further dredging.

Second, the whole delta ecosystem is inherently unstable and its hydrology poorly understood. The overall gradient of the delta is very low at only 1:3300. There are hundreds of interconnecting channels whose relative importance changes markedly from year to year: 'an extremely complex system of flow that is subject to changes in route and even direction' (Greenpeace 1991: 1). In a critical internal memorandum, Peter Smith (1989), an ecologist with the Department of Water Affairs, argued that historically the Boro River had not always been the main outflow from the delta and might not continue to be so (see also Wilson 1973); and that the proposed bunding channel might be unnecessary because in the years when extra water would be needed there was little flooding in the Boro floodplain. Smith noted that the estimates of flows used in the calibrations of the consultants' model differed from the actual recorded flows by amounts of the same order as the estimated increased outflow to be gained from dredging, hence it was within the boundaries of error that there would be no increased flow at all. Like Smith, Clift-Hill (1991) challenged the hydrological assumptions of the consultants' model, arguing that water availability without the scheme was generally greater than the conservative estimates generated by the model suggested. Hence there was no immediate need for the scheme. Smith agreed, arguing in an internal memorandum that there was no demonstrable need for the project although 'much of the data used ... assumes implementation of the scheme' (1989: 17). His conclusion that nothing should be done before an independent reassessment of the project is in line with the precautionary principle under which potentially harmful activities should be undertaken only if it can be demonstrated that there will be no environmental damage and, one might add, only if it can be demonstrated that the potential benefits will be realised.

The role of NGOs

While there was disagreement between scientific and technical experts on the economic viability of the SOIWDP, this alone was not the principal reason for the termination of the project. The decisive factor was an NGO campaign against the project. The role of NGOs and their contribution to the termination of the project are now examined.

Prior to Peter Smith's memorandum (noted above), the Kalahari Conservation Society (KCS) had expressed concern over the proposed Boro dredging in meetings with government. The conservationist lobby, particularly the Okavango Wild Life Society (OWLS) based in South Africa, were opposed to any development of the Okavango. However, as details of the proposed scheme were not made public until 1987, their opposition was not directed specifically against the SOIWDP.

KCS is the longest-established environmental NGO in Botswana and the only large national conservation organisation. Louis Nchindo, Head of Debswana, which runs the Orapa diamond mine, has been Chair of KCS since its inception in the 1970s. Debswana is a joint venture between the Botswana government and De Beers, the South African diamond giant. The 'independence and impartiality' of KCS has been called into question by more radical conservationists:

> [KCS] has always been closely associated with De Beers and its name, policies and activities conform remarkably well to advice given in a Report on Developing Tourism Revenue in Botswana. A few excerpts ... illustrate the grounds for suspicion: '*It is with this background of marketing psychology in mind that it is proposed that the Kalahari Conservation Club should be formed by the Government of Botswana to make use of the famous name of the Kalahari in order to establish Botswana as a high cost status symbol destination for international tourists. ... The Kalahari Club would, therefore, be fully and directly under the control of the Government of Botswana.*'
>
> (Williamson 1994: 24; emphasis added)

KCS has also been criticised in government circles for its white expatriate leadership and conservationist interests that may conflict with Botswana's development objectives. Nonetheless, KCS has a record of promoting conservationist ideals and has successfully lobbied for anti-poaching legislation. KCS stated its general position in a 1991 statement on the SOIWDP:

> It is not generally realized how much pressure KCS has brought to bear on Government over the last two years, nor the open and frank way that Government, particularly the Department of Water Affairs, has taken note and responded. ... By nature of its position in Botswana, KCS has decided not to adopt the more radical stance of external pressure groups such as Greenpeace since it would quickly lose credibility with Government and be much less effective.
>
> (KCS 1991: 4)

In the late 1980s, Eleanor Pattison, the Chief Executive Officer of KCS, was made a non-voting member of the government's Water Development Committees providing a route for KCS to state its views and enabling the government to claim it had consulted conservationists.

In 1988 KCS wrote to the Permanent Secretary of the Ministry of Mineral Resources and Water Affairs to suggest possible alternatives to the SOIWDP, such as a pipeline or canal from the delta. In May 1990 KCS and others attended a meeting to discuss their objections with the Department of Water Affairs (KCS 1991: 6). KCS had been reassured by the EIA and received further assurances that there would be no additional incursions into the delta. However, like Smith and Clift-Hill, they had doubts about the hydrological viability of the scheme. KCS offered to commission independent consultants to reassess the project.

Meanwhile the Department of Water Affairs hired an engineering firm to assist in technical details. Only when the general shape of the scheme was decided did the department engage local consultations through a series of small *kgotla* meetings in communities likely to be affected such as Maun. The *kgotla* is traditionally the local chief's court and the chief may call a meeting to discuss matters of importance in the community. However, most *kgotla* meetings are now called by chiefs at the request of government officials in order for the latter to explain policies which have already been decided. Although this is officially part of a 'consultation process', it is unusual for dissenting voices to be heard. However, in the series of *kgotla* meetings on the SOIWDP[1] some general, though muted, dissatisfaction was expressed. The term 'river improvement' was used in these consultations, giving the impression of limited works such as weed clearing. 'Dredging' was not mentioned, so that local people did not expect a large-scale project.

These local consultations occurred in parallel with equally low-key national consultations with KCS and others. The design, estimating and tendering process was taking place at the same time. However, no public announcement had been made about the SOIWDP since cabinet approval in late 1988. The Maun community was therefore taken by surprise when engineers from the South African company LTA arrived in November 1990.

The suspension of the project

The quickest to react were white residents, members of the so-called 'Duck Inn community', named after the meeting place in Maun favoured by tour operators and conservationists. Several of this group were members of the Maun branch of KCS. However, KCS refused to sponsor a public meeting at the Duck Inn in November 1990, although they were represented.[2] Questions were asked on why the Okavango had not been declared a World Heritage Site, on the EIA, on the loss of floodplains of importance as breeding sites for rare bird species, on the lack of interim provision for those who depended directly on Boro water while the works were in progress, on the danger of seepage due to river bed damage, on the negative effect on tourism in a supposed wilderness area, and on the possible extinction of certain fish species. The fear was expressed that the whole delta could be destroyed.[3]

Many of these concerns were expressed not by the white conservationists who had convened the meeting but by the Batawana. It was recognised that a campaign could only succeed if it was clearly demonstrated that opposition was widespread and not confined to a minority with interests in conservation and tourism. It was agreed to use official channels and contact the District Commissioner, the local tribal administration and central government departments to ask for a *kgotla* meeting. Although it was agreed that no organisation was being set up and everyone would attend the *kgotla* as individuals, an informal Batawana steering committee undertook to contact the authorities on behalf of the Maun community. Meanwhile Maun residents attempted to initiate an international campaign by contacting environmental NGOs in South Africa, USA, UK and elsewhere.

The position of KCS in relation to the activism in Maun is unclear. The South African press reported that KCS as a national organisation had, at least temporarily, stopped lobbying on the issue and had stated that:

> Now that the contract for the project has been awarded, KCS feels any further lobbying can have no practical purpose and that the best course of action is to continue to observe how the project develops.
> (Johannesburg *Sunday Times*, 2 December, 1990)

A white conservationist resident of Maun was also reported as claiming the KCS had 'been ordered by the government to keep its nose out of the issue'. However, the official KCS position is that they never issued a directive for the Maun branch to cease campaigning, although they cautioned them not to engage in activities which would conflict with KCS objectives in Gaborone. KCS believes that it retains more credibility with government if it avoids publicity.

At a second public meeting at the Duck Inn in December 1990 it was noted that De Beers were no longer involved, hence it could no longer be argued that water was required for 'national development'. Effectively the only reason for dredging was water for Maun, but as 'the meeting here is totally against dredging' the government should delay the project pending an independent investigation. One contribution from the floor summed up much of the feeling:

> I speak as a tribesman. You know we had a dredging scheme that failed. ... I saw the terrible effects. ... Dredging is destructive to the environment, and there is still no guarantee that you can drink the water. ... It is like saying you killed a snake by hitting it on the tail.

> Government can get water in some other way. ... Dredging is the last thing. Dredging is something we don't want.
>
> (*Questions asked at a Public Meeting held at the Duck Inn on Wednesday 12 December 1990*, mimeo)

Newspaper reporting in South Africa played a role in the evolution of the campaign. First, a newspaper reported that the Maun residents planned to sabotage construction (for example, Johannesburg *Saturday Star*, 8 December 1990). The Batawana steering group felt obliged to point out when writing to the Minister of Water Affairs on 7 December 1990 that 'we are not a fighting group' and 'we regret that the tone of the press is rather harsh'. Second, it was reported that Greenpeace International was 'considering launching an international campaign called "Diamonds are for Death" to save the swamps' (Johannesburg *Sunday Times*, 2 December 1990; also *Saturday Star*, 8 December 1990; *Sun*, 11 December 1990). The *Sunday Times* and the *Sun* also quoted the Environmental Investigation Agency (EIA) of London as saying: 'We fully support the protest groups inside and outside Botswana and are concerned about De Beers' involvement with the project.'

A *kgotla* meeting in January 1991 to debate the project lasted for over six hours. A high degree of mobilisation was achieved. Local businessmen had lent trucks to transport people to the meeting. Most of those present were traditional leaders and representatives of communities in and around the delta. The meeting was attended by Archie Mogwe, the Minister of Mineral Resources and Water Affairs, who was questioned by at least twenty speakers, none of whom supported the project. Mogwe appeared to have been surprised by the strength of feeling. Even the local MP from his own ruling Botswana Democratic Party, Bahite Temane, spoke against the project, and accused the government of not having told local people about the dredging. Other speakers appealed to the idea of democracy:

> The huge turnout of locals gave the lie to suggestions by proponents of the scheme that opposition to it was being organised purely by a band of expatriate whites for their own ends. ... Accusing the government of arrogance, ... ex-game ranger John Ben said Botswana was a democratic country and Minister Mogwe must understand that 'the Kgotla rules – he must come and speak to us here, and answer us here.'
>
> (Johannesburg *Saturday Star*, 12 January 1991)

Immediately after the meeting the Minister announced that the SOIWDP was suspended 'pending further consultation', while restating that 'Government remains satisfied that this project offers the best long-term solution to the water shortage in Maun ... and other settlements along the Boteti and Nhabe rivers'. The suggestion that there would be no serious environmental impact was reiterated. The reason given for the suspension was that 'certain members of the community still had reservations about the project'. In retrospect Moremi Sekwale, then the Director of Water Affairs, now says that the scheme was shelved as a response to threats from Greenpeace. However, at the time the government press release did not mention international opinion or Greenpeace.

The termination of the project

The Greenpeace visit took place from 29 January to 12 February 1991. The team comprised Allan Thornton of EIA (also a Greenpeace board member), Alison Ross, a wildlife campaigner with Greenpeace UK, and Tony Marriner, Head of Greenpeace Communications. To Alison Ross, the idea of a 'Diamonds are for Death' campaign was never a Greenpeace proposition. The suggestion that Greenpeace might campaign on this slogan was probably made by sympathetic South African conservationists.

The first two days were spent on introductory meetings with ministries in Gaborone before flying to Maun to meet what by now was the Tshomorelo Okavango Conservation Trust (TOCT) with a membership of white Maun conservationists and local Batawana leaders. TOCT had arranged meetings with local tribespeople and visits to sites in and around the delta. On the last day of the visit a joint press communiqué was issued by the Botswana government and Greenpeace International. It stated that Greenpeace had 'appealed to the government of Botswana to establish the Okavango Delta Ecosystem as a World Heritage Site' and to sign the Ramsar Convention on Wetlands of International Importance and that the Botswana government was considering this. The communiqué stated that the SOIWDP was suspended and that:

> The Botswana government will not dredge for the sake of dredging, but ... does have to keep the dredging option open, if in the end it is the only option through which Government can meet its obligation to provide water to Maun and communities down the river.
>
> (Joint Communiqué issued by Botswana Government and Greenpeace International on the SOIWDP, 12 February 1991)

Greenpeace then wrote to the President of Botswana, Quett Masire, urging that the dredging be abandoned, that the integrity of the delta ecosystem be protected and that an independent review be commissioned on the water needs in the area. Greenpeace's report on the visit (Greenpeace 1991) was circulated widely to international NGOs and the press. However, Greenpeace did not initiate a full international campaign on the issue.

The Botswana government, meanwhile, had commissioned an independent review by the World Conservation Union (IUCN). IUCN's team leader, Ted Scudder, an anthropologist from CalTech, assembled a team of thirteen people from four countries, constrained both by IUCN's policy of not recruiting its own personnel for such studies and by the government's stipulation that the team not include people who had recently worked in Botswana. The team of hydrologists, biologists, engineers and social scientists began work in October 1991 and presented its report in May 1992 (IUCN 1992). The recommendation to terminate the project was unanimous. Although aspects of the review team's work were not complete, in particular the hydrological modelling and the study of the international aspects of river basin management, it was felt that the weight of argument was sufficient for a clear recommendation. The team argued that insufficient attention had been given to alternatives, there had been inadequate analysis of some of the impacts on the natural resource base and that there were institutional weaknesses and design flaws in the Maun reservoir. It was also noted that:

> Analysis by the IUCN review team leads to the conclusion ... that the low income majority ... would, on balance, be worse off if the proposed scheme had gone ahead ... Numerous examples of overestimation of benefits and underestimation of costs further weaken[s] ... a project which is no longer justified on humanitarian grounds and which, as currently conceived, is more for the benefit of Maun and Orapa than for the rural majority.
>
> (IUCN 1992: 1–4)

The team analysed a 'preferred alternative' which would meet the objectives of the SOIWDP more sustainably and at lower cost. It suggested that Maun's water needs could be met indefinitely from the joint use of groundwater, surface water and a water treatment plant, and that the Orapa diamond mine could obtain enough water from groundwater sources. There were recommendations for improving the household economy of the riverine communities. These include building on the potential offered by water-related natural resources such as thatching grass, reeds, fish and wildlife, and integrating the village sector into the development of Maun and the tourism industry.

On the same day the IUCN released their report (21 May 1992), the government announced that the SOIWDP was 'terminated'. However, it was made clear that the cancellation was due to the extent of local opposition and not to the IUCN critique. The Director of Water Affairs, Moremi Sekwale, later claimed that had the review been undertaken on the basis of a contract direct with the government then it would not have been accepted and public release would not have been allowed. Scudder has criticised the Botswana government for failing to respond to the IUCN report. In particular he has suggested that the government may be deliberately 'punishing' Maun and the local riparian communities by not implementing or even appraising his team's 'preferred alternative' (Scudder 1994).

We have seen that there were two clear and apparently forced policy changes on the part of the Botswana government and its Department of

Water Affairs: first, to halt work on the project while an independent study was commissioned from the IUCN; second, to 'terminate' the project before responding to the report of the IUCN study team. However, while dropping the SOIWDP the government has made it clear that it considers the waters of the Okavango to be a national resource which it retains the right to develop as part of the modernisation of the country.

Analysing the policy change

There are factors which may partly explain the policy change that are peculiar to Botswana, such as the special role of the expatriate community in the country. These are not considered here. Instead this section concentrates on four explanations of more general application. They relate to the democratic attributes of the political context, the complementary role of different NGOs, the internationalisation of the issue, and the relative lack of impact on major economic stakeholders. We now examine these in turn.

The domestic political context

Botswana is generally regarded as one of Africa's few successful multi-party democracies. In order to test claims that the policy change over the SOIWDP resulted from the democratic nature of modern Botswana we suggest, following the approach in Chapter 1 of this volume, that democracy can be considered in terms of five aspects: accountability through multi-party elections; civil and political rights; diversity of power centres; limited reach of the state; and political participation. Taken together, these five features of democracy capture the complexity of political culture better than a scale measuring less or more 'democracy' or 'democratisation' as a uni-dimensional process.

First, with respect to accountability through multi-party elections, there has been clear and public espousal of the ideals of multi-party democracy. However, the Botswana Democratic Party has won every election since independence and in practice Botswana is a '*de facto* one party state through the ballot box' (Tsie 1993: 36). The second largest party, the Botswana National Front (BNF), provides a credible opposition and there is no real dispute that the elections have generally been 'free and fair'. However, there is little prospect of a change of government through the ballot box.

Second, with respect to civil and political rights, Botswana is in principle open and politically 'free'. However, up to 1989 at least it was claimed that 'formal and informal censorship pressures remain strong for reporting on the main opposition party, the BNF' (Zaffiro 1989: 67). The self-conscious adherence to liberal democratic ideals has meant that it has been possible to mount effective campaigns against abuses of human rights or personal freedoms, even on behalf of expatriate individuals. However, it is not clear how meaningful this political 'freedom' has been in terms of generating widespread discussion on important national issues.

Third, with respect to diversity of power centres, Botswana to some degree exhibits a classical separation of state from society. Both local councils and the judiciary have a degree of independence and power in their own spheres. Traditional leaders and structures such as the *kgotla* remain important at the local level thus contributing to an internal plurality of power. At the international level, the Botswana government is relatively independent of international agencies and donor community and is strong enough to treat them on its own terms, so that they do not represent such an important alternative source of power in Botswana compared to other African countries. However, foreign investors and companies, particularly those from South Africa, are powerful actors within their own industrial sectors.

Fourth, the strength and apparent democratic legitimacy of the Botswana state allows it to dominate areas such as rural development and education whereas in other African countries there is space for NGO activity. There is a tendency for the government to co-opt independent initiatives.

Fifth, with respect to political participation, traditional forms of local participation persist in Botswana although they are of decreasing importance. The government uses the local *kgotla* meetings to consult with or inform local communities about policies, and in principle they are also open to NGOs. However, there is no strong popular political culture at the national level. Voluntarism is still seen as cutting across traditional commitment to community, though this attitude is changing with urbanisation and the spread of commercialism and wage economy.

To summarise, Botswana is not a model democracy, though it does 'score' well in certain respects, particularly multi-party elections and a plurality of internal power centres. However the peaceful transition to independence meant democratic institutions were all 'handed down' and there was no growth of popular political participation. Attention now turns to the applicability of these five features of democracy with respect to the anti-SOIWDP campaign.

There was some reference to elections and the political party process during the campaign. One of the Batawana leaders opposed to the SOIWDP was Moutsamai Mpho, an ex-Independent MP. BDP MP Bahite Temane and other BDP members spoke alongside Mpho. There were reports of a possible electoral alliance between opposition parties that might include environmental issues in its platform (Johannesburg *Weekly Mail*, 11–17 January 1991). The government is generally sensitive to any issue that could become a focus of opposition during an election campaign. However, the campaigners generally played down any question of electoral opposition and emphasised instead the unity of the local community against the project.

The existence of alternative power centres was more important to the success of the campaign. The Duck Inn community, comprising conservationists and tour operators, could not be discounted given the importance of tourism to the economy. More important was the opposition of the Maun Council and traditional authorities such as the Chief of the Batawana. These relatively independent power centres articulated opposition to the SOIWDP. The traditional institution of the *kgotla*, though usually a vehicle for one-way communication from government to communities, was used to generate and demonstrate local opposition. The importance of these groups can be gleaned by a quote from the President in response to a question about water shortage in Maun in 1994:

> I suppose this is where democracy conflicts with development. We could have built a dam here a couple of years ago and there would be water galore but mainly because those who have interests in the tourism industry, they thought it was going to spoil the environment or bring in lots of people here to whet the appetite of government to build even more dams. So there was Green Peace [*sic*], there was the local people, they incited the local tourist operators, they incited the people in the Kgotla and they all refused. So in a democratic set up, people must make their beds and lie on it.
>
> (His Excellency Dr Quett Masire, President of Botswana, interviewed in the *Okavango Observer*, 12–18 July 1994)

Although there was a high degree of popular participation during the campaign, this was unusual. One cannot argue that the policy change arose because of any participative political culture, but rather that it occurred despite the closed nature of the policy process. The low-key representation on government committees by the KCS and the equally low-key consultations with communities before the decision to go ahead with dredging was well known does not constitute full political participation and did not contribute to the policy change.

With respect to civil and political rights, although there was no open suppression of the campaign, our research reveals hints of government pressure on KCS and veiled threats against expatriate campaigners. It is also interesting that press coverage of the campaign in Botswana was sparse, despite the supposed independence of the Botswana press, and that the campaign received most press coverage in South Africa.

In conclusion, the attribute of democracy that contributed most to the success of the campaign was the presence of alternative power centres, particularly at the local level. The government often claimed that the climb-down over the Okavango was proof of its democratic credentials. One opponent to the SOIWDP was prepared to give the government credit for backing down. In an interview with an author of this chapter, Peter Smith commented: 'You wouldn't find any other African government doing this.'

The complementary role of different NGOs

The NGOs involved in the campaign included local, national and international NGOs. The campaign was relatively unusual in Botswana where in general NGOs have little importance. There are important differences between NGOs in terms of their cultural background and historical development as movements (Chapter 1). In particular, western-style conservation NGOs must be distinguished from indigenous national and community-based NGOs. The former include KCS whose concerns are still sometimes dismissed as a 'white man's problem'. The recent modest growth in NGOs includes some local conservation NGOs, partly in response to international donor pressure. This carries the danger of a proliferation of 'environmental' NGOs with shallow commitments to community values. However, some of the new local NGOs, like TOCT, may combine local community and white conservationist interests in a genuine effort to represent local environmental concerns.

Although NGOs have been involved in formal consultations on environmental policy, they do not have a strong input. For example, Botswana's National Conservation Strategy (NCS) was adopted in 1990 after a substantial consultation programme including NGOs, although NGOs have not been able to exert any real pressure towards implementation of the NCS. However, there have been a few cases where NGOs have been able to prevent the implementation of certain policies. In Botswana, NGOs have tended to be restricted more to a reactive obstructive role rather than a proactive constructive one.

In the SOIWDP campaign NGOs of different types undertook different kinds of activities. KCS, for example, thought in terms of maintaining a long-term collaborative relationship with government and avoiding public confrontation. It was able to lobby behind the scenes and to obtain technical information. It is possible that its links with industry had some bearing on Debswana's withdrawal of support from the project.

The individuals who formed TOCT worked by mobilising the local population and were not afraid of confrontation. After the SOIWDP was shelved, TOCT continued to work on mass education in conservation issues. It should, however, be noted that in 1993 and 1994 TOCT suffered a split between safari operators and local people over the government's land-use plan. If a new project similar to the SOIWDP were to be planned, it is not certain that TOCT would again be able to mobilise.

The roles of international NGOs involved in the campaign were different again. The most openly confrontational were the South African activists including OWLS. Greenpeace brought a reputation more than anything. Mentioning their name provided a credible threat even though there is no evidence that Greenpeace made any decision to campaign beyond sending a study team, or made any formal threats of any kind. The independent study team lent scientific credibility to the campaign through IUCN's scientific research capacity.

The NGOs therefore performed different roles which together complemented each other effectively. They were not, however, deliberately coordinated. As we have seen, TOCT was formed through disagreement with KCS, and the two NGOs did not work together. TOCT did work with Greenpeace during the study tour, but there was no coordinated international or national campaign as such. NGO relationships were *ad hoc*, what Eccleston would call a *coalition* on a 'single event joint campaign ... among fairly diverse NGOs [with] division of labour [and] limited life' rather than a permanent *network* (1996: 74).

Internationalisation of the Issue

Another interpretation is that the internationalisation of the issue forced the government to change policy. Eccleston suggests that collaboration with northern NGOs can 'widen the political space within which [southern NGOs] campaign', although a concomitant danger is that government may perceive international collaboration as a threat, thus curtailing domestic political space even further (1996: 67).

Northern-based international NGOs have assumed increasing importance in Africa. However, Botswana's record of economic growth has protected the government from pressure to allow open access to northern-based NGOs. International NGOs have tended to work through government agencies rather than independently or in partnership with indigenous NGOs. Nonetheless, local branches of international NGOs do operate in Botswana.

In the case of the Okavango campaign, the involvement of international NGOs appears to have aided the campaign. International NGOs claimed legitimate interest by framing the Okavango as an international, rather than just a local or national, issue. The role of South African environmentalists in placing stories in the South African press cannot be discounted. Greenpeace and IUCN brought scientific and organisational resources to bear which were beyond the capacity of TOCT. The involvement of South African activists and Greenpeace presented the government with an external threat, real or perceived. To some extent the Botswana government also helped promote the broader view of the SOIWDP as an international issue by commissioning the IUCN to undertake the independent assessment. However, it remains to be seen whether summoning international assistance will cause a backlash for the relatively weak Botswana NGO movement over the longer term or whether it will continue to be a powerful occasional weapon.

The fact that other international actors were involved may also have been important. The decision of De Beers to withdraw support from the SOIWDP may have been made with De Beers' international reputation and interests in mind. There were other international actors as well, such as SMEC (part-funded by the Australian government) and potential donors who may have backed the project financially. Although their interests would have been enhanced by the project they might have been targets for lobbying if a full-scale international campaign had materialised.

The lack of impact on major economic stakeholders

We have seen how the campaigners, particularly those outside the country, tried to emphasise links between dredging the Boro and economic sectors of central importance to Botswana and its ruling elite, notably the diamond and beef industries. Both these sectors involved links with international actors on whom leverage could be exerted. Arguably this helped the campaign.

However, it is equally arguable that it was the relative *lack* of importance of the SOIWDP to the major economic stakeholders within Botswana that allowed the government to drop the project. According to this view, it was not that De Beers or Debswana were put off by threats, but rather that the Orapa mine had little need for additional water. Also the downturn in the diamond industry reduced the urgency for water at Orapa, so there was no point in De Beers risking its reputation with international environmentalists by backing the scheme.

Similarly, the efforts of South African environmentalists and others to depict Botswana's 'beef barons' as determined to take over the country for cattle ranching may have some truth in it. The Northern Wildlife Area was being divided into separate protected areas, with large areas of rangeland in between given over to cattle, such as that immediately to the north-east of the Okavango Delta. However, there is no evidence of any immediate plans to bring cattle into the delta itself, so there was no need for the beef lobby to back the SOIWDP.

Where the interests of major economic stakeholders are directly implicated there is less chance of forcing a policy change. In the case of the SOIWDP, even with the international links and possibilities of international lobbying, it is not clear that the campaigners could have prevailed if the project had been of significant economic importance to Botswana.

The limitations of the NGOs

On the whole the NGOs were successful in their immediate objective of stopping the dredging. However, they have not been able to persuade the government to join the Ramsar Convention or to adopt the IUCN team's alternative proposals for local development. Partly this is a question of a lack of continuing capacity on the part of the NGOs concerned. In different ways, they all responded to what they saw as a crisis by a big short-term effort.

For example, the individuals who formed TOCT and OWLS reacted to the crisis by investing a lot of time over a short period to mobilise the communities and to communicate to the media and international NGOs. This degree of voluntary effort could not be continued over a protracted period. Apart from some minor material assistance from OWLS and IUCN, TOCT had little organisational capacity. Greenpeace had never prioritised conservationist issues in Southern Africa, hence its capacity for long-term campaigning was extremely limited. IUCN was contracted to undertake a study and its team of professional scientists from different parts of the world could not be expected to pursue the issue once their report was completed, although Ted Scudder did so personally to some extent (Scudder 1994). The only NGO involved which has long-term capacity is KCS. It remains to be seen whether its quiet 'insider' strategy will gain results over the long term.

Apart from lack of capacity, another weakness was that the NGOs did not represent clearly united interests. One contention is that the anti-SOIWDP campaign was led by expatriates in Maun with interest specifically in wildlife conservation and connections to safari companies. In this view the local NGOs such as TOCT were not truly representative of local opinion and Greenpeace took up the issue without fully understanding the local situation. This possible lack of 'representativeness' and of full political accountability means that in the longer term the Botswana government may feel free to ignore these NGOs. It appears that while there may have been a temporary coincidence of objectives between the white conservationists and the members of the local communities in TOCT, there was no underlying unity of interests. Hence a concerted long-term campaign could not be expected.

Conclusions and implications

From the various conflicting interpretations and competing explanations, what can be concluded? What are the long-term implications for NGO strategy? First, it appears that confrontational tactics can work, at least in the short term, in a multi-party democracy like Botswana where there is an internal plurality of power centres and if the tactics are used only in response to occasional crises. Multi-partyism is meaningful as a source of accountability in Botswana, and the government is sensitive to the need to maintain its democratic credentials. The ruling party is also careful to avoid allowing issues to develop so that opponents can use them to political advantage. Both local councils and the *kgotla* system provide some degree of alternative internal power centres. If there is clear dissent from a government position at the local level then the government may give way in order to maintain control over the long-term political agenda.

However, there may also be arguments in favour of collaboration even where the participative political culture is weak, as it tends to be in

Botswana. Although there is only a limited record of successful NGO influence obtained by the 'insider' methods of KCS, there is clearly some potential for behind-the-scenes influence. In general the Botswana government is not lacking in technical expertise or administrative capacity, so there is little potential for NGO influence through filling gaps. However, the environment may be one important area where the expertise built up to date is not sufficiently relevant. The Botswana government may be forced again to call on NGOs to assist, as it did with IUCN.

With respect to international links, there seems to be some danger of a backlash if NGOs operate without regard for the sensitivity of government to sovereignty and the desire to maintain control over national resources. If the issue can be defined so that it is clearly an international issue, this will help to avoid such a backlash. The Okavango basin is now subject to international negotiations between Botswana, Namibia and Angola. It is difficult for the government to continue to maintain that the Okavango remains purely an internal issue, even though the delta is still not nominated as a Ramsar or World Heritage Site.

There is also some evidence of a more positive long-term effect that may override the danger of a backlash. This is a sort of demonstration effect. The SOIWDP shows that it is possible to oppose government and win. For some time there have been occasional cases where NGOs in Botswana have been able to prevent certain projects or policies from going ahead. The more they do this, the more their credibility grows. It may not be long before Molutsi and Holm's (1990) verdict that civil society is weak in Botswana has to be revised.

This trend may combine with the increasing difficulty for the Botswana government in maintaining complete sovereignty over issues that other actors consider to be international, possibly opening up more political space for NGOs in Botswana. There may also be some potential for NGOs to refer to the government's international commitments. For example, Botswana has signed the Convention to Combat Desertification, which includes a commitment to local participation and the involvement of NGOs at all levels (Chapter 3). Thus, although confronting government carries the danger of reaction against the NGOs concerned, this may be outweighed by the worldwide trend towards an enhanced role for NGOs.

Notes

1 Our source here is notes taken at the *kgotla* meetings by a District Officer, Lejo van der Heiden.

2 The Okavango issue led to tensions within KCS, and a few months later most of the KCS Maun members split from KCS.

3 The meeting was minuted by Karen Ross, then voluntary secretary of the Maun branch of KCS and the author of a book on the Okavango (Ross 1987).

References

Clift-Hill, A. (1991) 'Comments on the Draft Final Report Southern Okavango Integrated Water Development: Sukwane Dam and Pipeline Feasibility/ Preliminary Design Study, SMEC', Maun: Department of Agricultural Research (mimeo).

Colby, M. (1990) *Environmental Management in Development: The Evolution of Paradigms, World Bank Discussion Paper No. 80*, Washington DC: World Bank.

Eccleston, B. (1996) 'Does North–South collaboration enhance NGO influence on deforestation policies in Malaysia and Indonesia?', *Journal of Commonwealth and Comparative Politics*, XXXIV, 1: 66–89.

Greenpeace (1991) *Okavango – Delta or Desert? A Question of Water*, London: Greenpeace.

IUCN (1992) *The IUCN Review of the Southern Okavango Integrated Water Development Project: Draft Final Report*, Gaborone: World Conservation Union (IUCN).

Hyman, R. (1974) *The Failure of South African Expansion*, London: Macmillan.

KCS (1991) 'Okavango water development – a review of KCS input', *KCS Newsletter*, 31: 4–6.

Lomba, R. (1992) *Conservation and Development: Botswana's Cattle Experience*, Johannesburg: Okavango Wild Life Society.

Molutsi, P. and Holm, J. (1990) 'Developing democracy when civil society is weak', *African Affairs*, 89, 356: 323–40.

Moyo, S., O'Keefe, P., and Sill, M. (1993) *The Southern African Environment: Profiles of SADC Countries*, London: Earthscan.

Porter, J. (1989) 'Response to Mr. P. Smith's Criticisms of July 1989: Southern Okavango Integrated Water Development', Albury, Australia, 19 July (mimeo).

Ross, K. (1987) *Okavango: Jewel of the Kalahari*, London: BBC Books.

Scudder, T. (1994) 'Is government punishing Maun?', *Okavango Observer*, 14 October: 1 & 5.

Skinner, J. (1989) 'Review of Environmental Impact Studies concerning the Southern Okavango Development Project, Botswana: Final Report' (mimeo).

SMEC (1987a) *Southern Okavango Integrated Water Development Phase I: Final Report – Summary*, Cooma, Australia: Snowy Mountains Engineering Corporation.

SMEC (1987b) *Southern Okavango Integrated Water Development Environmental Impact Study: Final Report – Volumes I, II and III*, Cooma, Australia: Snowy Mountains Engineering Corporation.

SMEC (1988) *Southern Okavango Integrated Water Development Inception Review Report – Detailed Design: Maun Reservoir and Associated Works*, Cooma, Australia: Snowy Mountains Engineering Corporation.

Smith, P. (1989) 'The Southern Okavango Integrated Water Development (SOIWD) Project' (mimeo).

Tsie, B. (1993) 'The political context of Botswana's development performance', *Southern Africa Political and Economic Monthly*, 6, 12: 35–9.

Van de Laar, A. (1996) 'Economic theory and the natural environment: a historical overview', *ISS Working Paper Series 209*, The Hague: Institute of Social Studies.

Williamson, D. (1994) *Botswana – Environmental Policies under Scrutiny: The Lomba Archives*, Cape Town: Lindlife.

Wilson, B. (1973) 'Some natural and man-made changes in the channels of the Okavango Delta', *Botswana Notes and Records*, 5: 132–53.

Zaffiro, J. (1989) 'The press and political opposition in an African democracy: the case of Botswana', *Journal of Commonwealth and Comparative Politics*, XXVII, 1: 51–73.

8 An NGO campaign to protect the Hadejia-Nguru wetlands in Northern Nigeria

Susan Carr and Obasi Ogbonnaya

This chapter concerns the combined efforts of two linked non-governmental organisations (NGOs), namely the Nigerian Conservation Foundation (NCF) and the Hadejia-Nguru Wetlands Conservation Project (HNWCP), to protect a wetlands area in northern Nigeria. An important aspect of these efforts was a campaign to prevent the construction, or at least reduce the impact, of a major new dam at Kafin Zaki that threatened the existence of the wetlands.

At first sight the policy issue and the political context appeared to set a daunting challenge for non-governmental organisations attempting to influence government policy decisions. The literature on the factors that contribute to NGO influence suggests that success is more likely when NGO objectives do not conflict with those of powerful vested interests. It also suggests that NGO influence is more likely under democratic conditions, where citizens are free to gather and to lobby government, and government is more likely to be accountable and responsive to citizens' demands (Potter 1996: 10). In this case, the proposed dam formed part of a long-standing government programme to irrigate an area prone to drought and to help feed Nigeria's burgeoning population. Commercial as well as political interests were involved and stood to gain, while the losers were people downstream who depended on the wetlands for their livelihood, but who had little political influence. The potential leverage of an electorate could not be exploited, since Nigeria had had a military government, not a democratically elected one, for most of the period since the NCF and the HNWCP were established.

Despite these apparently considerable obstacles, the NCF and the HNWCP did achieve some notable successes in protecting the wetlands from the dam construction programme, especially during the most intensive part of their campaign between 1992 and 1993. In March 1993, shortly before work on a third major dam was due to recommence, NCF were granted an audience with Chief Shonekan, head of the Interim Civilian Government, at which they pressed their case for a full and independent environmental impact assessment of the proposed dam. In the following month, HNWCP held a policy workshop on the management of the river basin that was attended by key decision makers, including officials from the Federal Ministry of Agriculture, Water Resources and Rural Development.

Subsequently the wetlands project staff were regularly consulted about water releases from upstream dams to the wetlands. In late 1993 the government cancelled the contract to build the dam and eventually it shelved the dam-building programme. The main reason was the deteriorating economic and political situation at that time. Nevertheless, the scientific and economic case made against further dam construction by HNWCP and NCF may have contributed to the decision.

This chapter describes the key features of the Nigerian Conservation Foundation, the Hadejia-Nguru Wetlands Conservation Project, and the Nigerian government's programme of dam construction. It examines the strategies and tactics that the NCF and the HNWCP adopted to protect the wetlands in order to establish how they avoided, circumvented or overcame political obstacles and achieved the success that they did. It distinguishes between the factors that may have general application and those that are peculiar to the Nigerian situation.

The two NGOs

The Nigerian Conservation Foundation

The Nigerian Conservation Foundation is the most prominent environmental NGO in Nigeria. It was founded in 1980 and registered as a non-governmental charitable trust with the Federal Ministry of Internal Affairs in 1982. In 1989, it became Africa's first associate member of the World Wide Fund for Nature (WWF). Its membership is mainly Nigerian although a few expatriates are also involved. It is based in Lagos and most of its 2,000 or so members are from that area. Its committees and governing body are drawn from the elite of Nigerian society, including chiefs, former politicians and ministers, prominent industrialists and academics. Many have direct lines of communication to, and move in the same social circles as, government ministers and senior civil servants.

NCF has played a significant role in the development of environmental policies to protect wildlife in Nigeria. It has been instrumental in the establishment of a national parks policy, the development of the Federal Environmental Protection Agency, the enactment of a decree to protect endangered species, the drafting of the government's National Conservation Strategy, the introduction of legislation on environmental impact assessment, and the encouragement of environmental education in schools.

One of the ways in which NCF furthers wildlife conservation in Nigeria is to identify and establish conservation projects in valuable wildlife areas. NCF seeks the funding for projects, or acts as a Nigerian partner for international organisations wishing to fund projects, but once the projects are

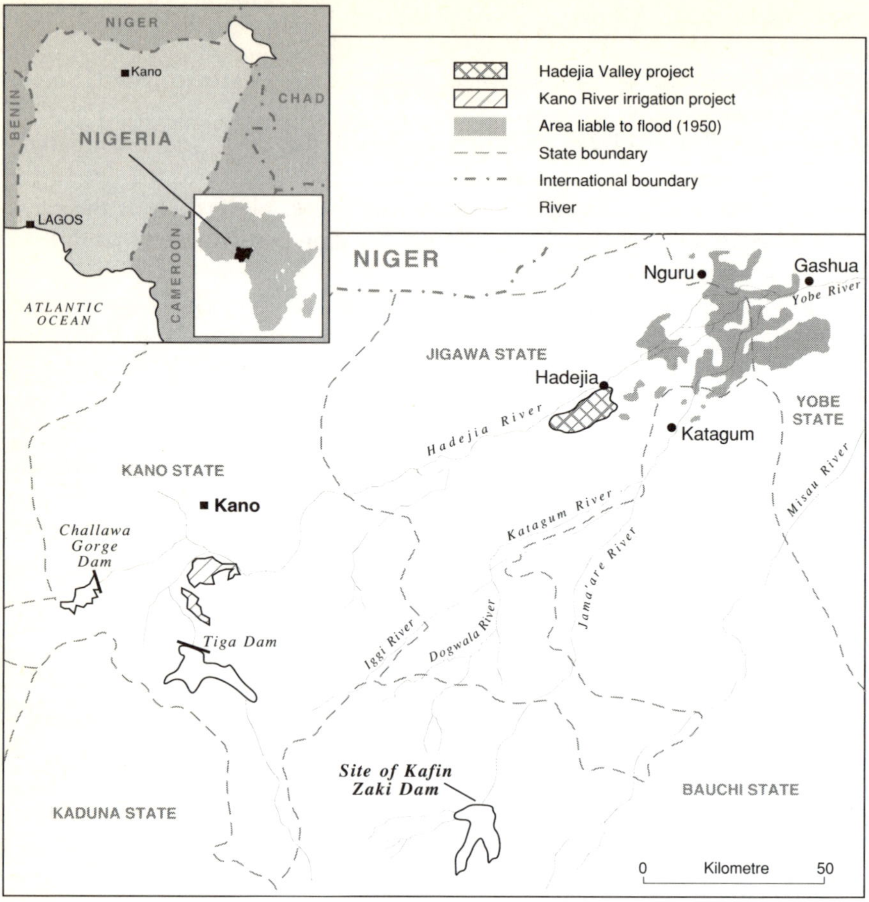

Figure 8.1 The Hadejia-Jama'are river basin.

established they are generally managed at 'arms length', with NCF having limited further involvement. Amongst the earliest of such projects was one in the Hadejia-Nguru wetlands.

The Hadejia-Nguru Wetlands Conservation Project

The Hadejia-Nguru wetlands are situated in the far north-east of Nigeria, in an area bordering the Sahara Desert that is otherwise arid or semi-arid. The wetlands are fed by floodwater from the Jama'are and Hadejia Rivers, where they join to form the River Yobe, which subsequently flows through Borno State into Lake Chad (Figure 8.1). The wetlands form where the Jama'are and Hadejia Rivers cross ancient sand-dunes, which cause the water to disperse into multiple river channels, producing a complex pattern of permanently and seasonally-flooded land and dry areas (Hollis et al. 1993: 5). Flooding reaches its greatest extent towards the end of the rainy season, in August and September. Subsequently the floodwaters recede, leaving isolated pools. In the dry season, between October and April, the main rivers may cease to flow at all.

About 1 million people live in the wetlands area, deriving their livelihoods mainly from fishing, cattle grazing, wet season rice farming and flood recession agriculture. The extent of the wetlands has been affected by drought and by the construction of two major dams, the Tiga Dam and the Challawa Gorge Dam, which have impounded key tributaries of the Hadejia River. At one time, in the 1950s and 1960s, the flooded area used to cover 2,500–3,000 km^2 during the rainy season, but surveys in the late 1980s and 1990s have shown a much reduced flooded area of between 500 and 1,000 km^2 (Hollis et al. 1993: 39, 67; Barbier and Thompson 1998).

The wetlands are well-known nationally and internationally among conservationists for their wealth of bird life, especially waterfowl such as ducks and wading birds. In terms of wildlife value they rank among the world's most important wetlands and would qualify as a wetlands of international importance under the Ramsar Convention if Nigeria were a signatory of this intergovernmental treaty.

In 1985, increasing concern about the decline in the extent of the wetlands, highlighted by a particularly severe drought in 1984, prompted the IUCN (World Conservation Union) to work with the Nigerian Conservation Foundation to initiate a project in the Hadejia-Nguru area. The first phase of the project (1989–92) was supported by the UK-based Royal Society for the Protection of Birds (RSPB) and Birdlife International (formerly the International Council for Bird Preservation) because of the significance of the area for migrating birds. The area forms one of a line of wetlands south of the Sahara that provide a reception area for birds recovering from their 60–70 hour unbroken flight across the desert from Europe. The IUCN later helped secure funding from Finland (the Finnish International Development Agency and the Finnish Association for Nature Protection) and from the European Union.

The main aim of the wetlands conservation project was to maintain the wetlands ecosystem through regular flooding of the floodplain. Although the initial stimulus for the project was to protect wildlife, especially birds, it soon became apparent that this could not be the sole aim in an area where many people depended on the wetlands for their livelihoods. By 1990, three years after the project was established, the emphasis had changed from wildlife to water resources. The scope was broadened to encompass three main objectives:

1 to investigate and promote the sustainable use of the wetland resources for the benefit of both human and wildlife communities;

2 to explore resource exploitation alternatives aimed at optimising traditional land use practices, especially by developing demonstration projects in collaboration with local communities;
3 to work with both the Nigerian administration and the international agencies to ensure, through co-operative programmes, seminars and workshops, that planned developments represent a sustainable use of available resources.

(Source: HNWCP 1993; Hollis et al. 1993.)

The project operated locally through NCF, as the Nigerian counterpart to IUCN. From 1991, the IUCN assumed responsibility for coordinating the project. In addition, the project staff received guidance from advisory and management committees, which included representatives from ministries in the northern states of Borno, Yobe, Jigawa and Kano, the Federal Department of Forestry, the IUCN, RSPB and NCF.

The work of the paid project staff was augmented from time to time by student volunteers, research students and researchers from universities in the UK (University College London, the University of Cambridge and Cranfield University's Silsoe College) and the Netherlands (Wageningen Agricultural University) with previously-established links with northern Nigeria. Together they gathered biological information on wildlife, hydrological information on seasonal river flow and the variation in the extent and timing of flooding, and sociological information on the structure of local communities and their sources of livelihood. They also conducted farming systems research to look at the agronomy and land-use systems. They established a number of small-scale demonstration projects to encourage sustainable resource use in the wetlands, for example on community agroforestry, improved *fadama* (wetlands) rice production, improved beekeeping, donkey traction and fuel-efficient stoves. They promoted awareness of the need for the conservation of the wetlands among the local communities. The senior staff worked with middle-level technical staff in state ministries, providing training, supported by funding from the British Council. By this means the project staff hoped to influence more senior government officials indirectly on matters affecting the wetlands.

Although the wetlands project was initiated at a time when it was generally believed the Kafin Zaki dam would not be built, the detailed information gathered at the time subsequently proved to be of central importance in making the case against further dam construction.

Dam construction in the Hadejia-Jama'are river basin

The original proposals for large-scale dam construction and irrigation schemes in the Hadejia-Jama'are river basin upstream of the wetlands were based on the advice of international agencies such as the World Bank and the United States Agency for International Development (USAID) in the late 1960s to provide security from drought and thus ensure adequate food supplies for a rapidly expanding population (HNWCP and NIPSS 1993: 19). The Tiga Dam, completed in 1974, was the largest earth dam in Africa at that time, with a storage capacity of 1,845 million cubic metres (Hollis et al. 1993: 151). Constructed across a tributary of the Hadejia River, it was intended to supply water for the Kano River Irrigation Project and for Kano City. Nine other smaller dams, with a total storage capacity of 183 million cubic metres, were constructed in the Hadejia-Jama'are basin, most of them on tributaries of the Hadejia River.

The economic downturn in Nigeria in the 1980s brought the dam-building programme to a halt. However, a rise in oil prices as a result of the Iran–Iraq war generated the revenue to enable the Challawa Gorge Dam, with a storage capacity of 904 million cubic metres, to be completed in 1992, and a barrage to be constructed across the river just above Hadejia to provide short-term water storage for another irrigation scheme, the Hadejia Valley Project. Plans for the construction of a third large dam with a storage capacity of 2,585 million cubic metres, at Kafin Zaki on the Jama'are River, were revived at the same time (Figure 8.1).

Over the intervening years there had been growing national and international criticism of large-scale dam construction and irrigation schemes. For example, literature on large-scale irrigation development in northern Nigeria highlights problems of slow development, high capital costs, poor yields, underestimation of the volume and value of existing production, poor water supply and management, high costs of resettlement and discontent among the resettled, high management costs, and poor returns for irrigating farmers, particularly small farmers (Hollis et al. 1993: 147). More generally, such schemes have been criticised for encouraging expropriation of land by the state, capitalisation of holdings by a few rich farmers, inequalities as a result of unequal access to inputs, finance and other means of production, and the emergence of absentee landowners (Olofin 1993).

In response to this kind of criticism, in the late 1980s the World Bank proposed and aided the establishment of agricultural development projects in northern Nigeria to improve the productivity of smallholders' traditional irrigation practices. Handlifting of water from pits and ponds using a *shaduf* (a bucket lowered and raised by means of a counterpoised rod, like a seesaw) was replaced or augmented by small fuel-driven pumps and tubewells.

These dams and irrigation schemes, particularly the large-scale ones, severely affected the extent of flooding in the Hadejia-Nguru wetlands. Traditional irrigation in the wetlands relies on the extensive flooding of the area that occurs in the rainy season. As the floodwaters recede, rice is planted in the wettest areas, while a wide range of market garden crops is planted on the slightly higher adjacent ground. As the ground dries out, farmers keep their crops watered by lifting water from the remaining pools and streams. Fish stranded in the pools provide a good living for fishermen. Semi-nomadic pastoralists keep livestock on the uncultivable ground. The decline in the maximum extent of the flooded area, from around 3,000 km^2 in the 1950s to less than 1,000 km^2 in the 1990s, has led to conflicts between farmers and pastoralists over the reduced wet areas and migration to urban areas as sources of livelihood in the wetlands have disappeared.

In order to make a case for protecting the wetlands from further damaging developments upstream, the Hadejia-Nguru Wetlands Conservation Project team, with the help of advice from their sponsors and international researchers, set out to establish the economic benefits of the wetlands compared with those of the main large-scale irrigation scheme upstream, the Kano River Irrigation Project. This initiative built on work in the USA on ways of assigning a monetary value to the functions of wetlands so as to add weight to conservation arguments based on wildlife and aesthetic criteria alone (Adamus 1983, cited in Hollis and Bedding 1994).

The analysis showed that the produce from the wetlands (from agriculture, fishing and fuelwood) had a value (at 1989/90 prices) of between US$ 34 and 51 per hectare, compared with the value of crop production from the irrigation project of between US$ 20 and 31 per hectare (Barbier et al. 1993; Barbier and Thompson 1998: 435). Comparisons based on the value of production per cubic metre of water used were even more favourable to the wetlands: between US$ 9,648 and 14,548 for the wetlands compared with US$ 26 and 40 for the irrigation project. Additional wetland benefits, such as those from livestock and grazing, non-timber forest products, groundwater recharge, refuge for human populations during times of drought, and habitat for birds, were noted but not quantified in the researchers' report.

To extend this analysis, researchers working with the wetlands project from University College London gathered data on the water regime of the Hadejia-Jama'are river basin. The data included the volume of water entering the floodplain from upstream tributaries, the volume leaving the wetlands at Gashua, local rainfall and evaporation, soil moisture recharge and infiltration to groundwater. Using these data the researchers constructed a computer model that could be used to predict the extent of flooding in the wetlands (Hollis et al. 1993; Thompson and Hollis 1995; Barbier and Thompson 1998). The model was calibrated as far as possible against the known extent of flooding as recorded in satellite images and aerial photographs, and by aerial and ground surveys. It was then used to predict the likely effect on the extent of flooding month-by-month of a number of different irrigation development scenarios, compared with a baseline scenario that assumed a natural flow uninterrupted by dams and major irrigation schemes. For each scenario, the costs and benefits were calculated using the values for wetland or irrigated production estimated previously.

The scenarios included the baseline, existing schemes and potential policy options:

- Scenario 1 – baseline (no dams or large-scale irrigation schemes)
- Scenario 2 – Tiga Dam, Kano River Irrigation Project (KRIP) extended to 27,000 ha (the original plan for the first phase of the two-phase project), no release of water to the wetlands
- Scenario 3 – Tiga Dam, KRIP at existing area of 14,000 ha, releases of water to the wetlands in August
- Scenario 4 – Tiga Dam and KRIP as in Scenario 2, Challawa Gorge Dam added with year-round water releases for use along the Hadejia River, small dams on tributaries (simulated by reducing the flow from these tributaries by 25 per cent)
- Scenario 5 – full implementation of all four existing and proposed water-resource schemes (Tiga Dam and KRIP at 27,000 ha, the Challawa Gorge Dam, the Hadejia Valley Project, and the Kafin Zaki Dam)
- Scenario 6 – all four water-resource schemes as in Scenario 5, but with water releases to the wetlands in an attempt to provide a regulated flood of water to the wetlands while meeting the needs of current irrigation schemes.

(Barbier and Thompson 1998: 436–7)

According to the assumptions built into the wetlands model and into the calculation of losses and benefits from wetlands and irrigated production, Scenarios 2 to 6 would each result in a net loss in the value of production from the river basin overall, ranging from US$ 2 million (Scenario 3 – full implementation of the first phase of the KRIP scheme; 1989/90 prices) to US$ 20 million (Scenario 5 – full implementation of all four existing and proposed schemes). The gains from irrigated production would amount to 17 per cent of the lost wetlands production at the most (Scenario 2). Overall production losses could be reduced by releasing some water from the dams to provide a regulated flood in the wetlands (Scenario 3 rather than Scenario 2, and Scenario 6 rather than Scenario 5).

These detailed hydrological and economic analyses provided the Hadejia-Nguru Wetlands Conservation Project and the Nigerian Conservation Foundation with an invaluable resource to argue their case for protecting the wetlands when the government revived its plans to construct the Kafin Zaki Dam in 1992.

The campaign to protect the wetlands from the Kafin Zaki Dam

The revival of plans to construct the dam alarmed the NGOs associated with the wetlands, not only because it was intended to be much larger than any previous dam, but also because it would be sited on the Jama'are River, which until then had been almost free of restriction by dams.

News that the contract to construct the Kafin Zaki Dam had been re-awarded first reached NCF from staff on the wetlands project, who had been carefully cultivating good relations with local and regional government officials as well as with the engineers in the prospective construction company. NCF immediately decided to take action, as recorded in the minutes of its council meeting of 22 April 1992:

> the indiscriminate construction of dams in the NCF's HNWCP project (upstream) is of serious concern to the committee. The committee has therefore decided to do everything within its resources to call the attention of the appropriate authorities to the threats posed to the entire wetlands if sufficient water is not released into the Hadejia river system.

For the most part, the campaign was a low-profile one. NCF did not make much use of the media or attempt to raise public support, at least in the early stages of the campaign. In general, NCF did not wait to cultivate public opinion before taking up an issue. Instead it assumed that it was acting in the public interest: 'NCF has a vision which it believes will benefit the public even if the public does not think so now. The damage may be done while you wait to educate the public, although we do not rule out the need for public support' (Interview, June 1995, NCF official). Nor did HNWCP or NCF directly confront government or the dam construction company (Julius Berger Nigeria plc) about the issue. Confrontation was not a strategy favoured by HNWCP or NCF, except as a last resort. Instead they waged a powerful, politically astute and sustained campaign of persuasion behind the scenes, making full use of all their contacts in high places and the wetlands project's scientific expertise.

An eminent NCF council member (Chief Asiodu, a former permanent secretary in the Federal Ministry of Petroleum, and Secretary for Petroleum in the 1993 interim civilian government) subsequently summarised NCF's general strategy as being to: identify who has to take the decision; identify the vested interests and their arguments; counter those arguments; and then invoke the national patriotic interest (Interview, June 1995). In the case of the Kafin Zaki Dam, it was assumed that the decision would be taken at the highest level (by the head of state), with advice from the Federal Ministry of Agriculture, Water Resources and Rural Development (in particular, the department dealing with dams and reservoirs). Those with the strongest interest in the dam's construction would be the contractor (Berger), any politicians or ministry officials with a stake in the contract, and the main beneficiaries of the dam-building programme in Kano and Bauchi States. The arguments they would use would be the potential economic benefits of irrigation and, in the case of Bauchi State, the need to ensure an adequate water supply for Bauchi city. The NCF could counter these arguments by using their own analysis to show that the economic benefits of the wetlands far outweighed those of the irrigation schemes. Among the other arguments used by some NCF council members were that there were alternative means available to supply water to Bauchi city and, more speculatively, that the wetlands had the potential to generate income from eco-tourism in the future.

Initially the actions of HNWCP and NCF were relatively low key. The scientific information from the wetlands project was published and made available to government. Formal representations were made to government by correspondence, asking for an independent environmental impact assessment before construction began. International organisations such as the World Bank and Birdlife International wrote to the Nigerian Government in support of HNWCP's and NCF's view.

In 1993, with dam construction due to start in October, the NCF stepped up its campaign. Taking advantage of one of the military government's moves towards the election of a democratic government – the establishment of a Transitional Council – NCF secured a meeting with the Head of the Transitional Council, Chief Shonekan, who had been a member of NCF before he was co-opted into General Babangida's military administration. As a member of the NCF delegation later remarked: 'We all knew Chief Shonekan. All the members [of the delegation] were people Shonekan could not refuse [to see] anyway' (Interview, June 1995).

At the meeting, the NCF delegation raised a number of conservation-related issues, among them their concerns about the Kafin Zaki Dam (NCF 1993). They urged the government to reappraise the project by means of an independent environmental impact assessment. They mentioned that the cost of breaching the construction contract if the reappraisal were to be unfavourable would be less than the cost of lost production if the scheme went ahead. They put across their view that the overall impact, including that on downstream communities, should be considered. In response, the Secretary of State for Agriculture strongly defended the need for dam construction to go ahead, arguing that if the area was to feed itself after the year 2000 irrigated agriculture was essential. However, he proposed the establishment of an inter-governmental commission to coordinate the release of water from the dams for downstream users. Shonekan agreed, adding that NCF should be represented on the commission. Shonekan also agreed to consider the need for an independent environmental assessment once government had reviewed the wetlands project's findings.

Soon after the meeting, Chief Shonekan became the head of the government briefly when General Babangida was forced to hand over power to a caretaker civilian government by the public reaction that followed his decision to annul the results of the 1993 national election. The timing of key events surrounding the proposed construction of the Kafin Zaki Dam,

Table 8.1 Timing of key events surrounding the proposed construction of the Kafin Zaki Dam, in relation to the political regime and HNWCP/NCF's actions

	Kafin-Zaki Dam	Political regime	HNWCP/NCF's actions
1980		Civilian government: President Shagari	NCF founded
1982	Contract awarded to Dywidag Nigeria Ltd (advised by Diyam Consultants; supported by Dycherhoff and Widmann AG of Munich)		NCF registered
1983		Military coup (Dec): HoS Maj. General Buhari	
1985		Military coup (Aug.): HoS General Babangida	HNWCP initiated
1987			HNWCP formally established
1989			First major report on the wetlands distributed (Adams and Hollis 1988)
1992	Contract awarded to Julius Berger Nigeria plc (parent company German-Bilfinger and Berger)		Campaign against dam initiated
1993	Construction due to start October 1993. Suspended late 1993 pending initial payment to Berger by the Federal Government	Babangida annuls election results (June)	NCF delegation meets with Chief Shonekan (March)
		Interim civilian government headed by Chief Shonekan (Aug–Oct)	HNWCP-NIPSS workshop on management of water resources (April)
		Miltary coup (Nov): HoS General Abacha	Book on Hadejia-Nguru wetlands published (Hollis et al. 1993)
1994	Contract cancelled and judicial commission of inquiry set up (Feb)		

Key
HNWCP – Hadejia-Nguru Wetlands Conservation Project
HoS – Head of State
NCF – Nigerian Conservation Foundation
NIPSS – National Institute for Policy and Strategic Studies

in relation to the political regime and HNWCP's and NCF's actions, is shown in Table 8.1.

In parallel with NCF's campaign, the wetlands project team stepped up their own advocacy efforts, helped by the fact that their director at the time was from a highly respected political family. Their efforts culminated in a policy workshop (in April 1993) to discuss the management of the water resources of the whole river basin, hosted jointly by the HNWCP and the National Institute for Policy and Strategic Studies, a government think tank with access to General Babangida. The workshop was chaired by the chairman of the Federal Environmental Protection Agency, who as a former Minister of Agriculture had been closely involved in the original dam-building programme. Other participants included representatives from the Federal Ministry of Agriculture, Water Resources and Rural Development, the Hadejia-Jama'are River Basin Development Authority, state governments, dam and irrigation consultants, the Natural Resources Conservation Council, the IUCN, the World Bank (at their own request) and the North East Arid Zone Development Programme.

Every detail of the workshop was carefully planned ahead by the wetlands project team. They ensured that key decision makers were accommodated at the best hotel, where they might be expected to spend the evening socialising together, while project staff stayed in more modest accommodation and spent the evening making sure their presentations and plans for the workshop were as persuasive as possible (Interview, April 1994, former HNWCP staff member). Likely supporters of HNWCP's position were circulated with a discussion paper before the meeting, while likely critics received papers at the last minute. The project staff drafted a list of recommendations for the participants to review and to endorse, improve or delete as the meeting's conclusions. During the meeting, the wetlands project's consultants from University College London (Ted Hollis and Julian Thompson) demonstrated their hydrological model and made a forceful case for protecting the wetlands.

While the text and order of the draft recommendations were altered by the workshop participants, the basic proposals remained substantially as the wetlands project staff had intended. For example, the recommendations stated that the Tiga, Challawa Gorge and Kafin Zaki dams and the Hadejia barrage should be operated to satisfy the requirements of both upstream and downstream communities. Water should be released from the dams in the wet season to flood the wetlands to maintain their productivity and other functions. An environmental impact assessment and a cost–benefit analysis of the Kafin Zaki Dam should be undertaken and its design reassessed to ensure there were adequate outlets to provide artificial flood releases. A coordinating council should be established to formulate an integrated policy for the whole river basin. Facilities for data collection, processing, storage and dissemination should be strengthened. The workshop's recommendations were presented by the National Institute for Policy and Strategic Studies to General Babangida.

A direct outcome of the workshop and NCF's representations to

Shonekan was that the wetlands project staff were subsequently regularly consulted by the Hadejia-Jama'are River Basin Development Authority about releases of water from the existing dams to the wetlands.

In late 1993, construction of the Kafin Zaki Dam was suspended, just as it was due to begin. For whatever reason, the government failed to hand over the initial payment to the construction company. In early 1994, the contract was suddenly cancelled, without explanation, and a judicial commission of inquiry established to investigate financial irregularities in the award of the contract. It is difficult to disentangle the causes of the cancellation from the turbulent political context. General Abacha had taken over from Babangida and Shonekan in a military coup in the previous November. Although it is likely that the deteriorating political and economic situation was the main reason for the contract being cancelled, it is possible that increasing opposition to the dam, led by the well-supported case made by the NCF and the wetlands project, at least played some part in the justification for the decision.

Cynics suggest that the contract and its cancellation almost certainly involved the 'Nigerian factor', defined by Olofin as 'a heavy load of constraints such as corruption, nepotism, sectionalism, bribery, falsehood and so on, all of which render the "truth" useless and dangerous, and make project-implementors pursue the unreal' (Olofin 1993: 40). Babangida was on the board of the construction company Berger. Abacha was not. Newspaper reports of the dam inquiry referred to the 'contract upon contract imbroglio' (Babadoko 1995).

Abacha was generally viewed as sterner than Babangida, less willing to listen to advice and less amenable to reasoning and compromise. Under Abacha's regime, NCF and HNWCP, like everyone else, found it more difficult to exert their influence. NCF's direct links to government were weakened. A follow-up to HNWCP's policy workshop, in January 1995, was not so well attended. Ministry officials who had been due to attend withdrew at short notice. The dam issue had become politically sensitive and 'everyone in the water business [was] trying to keep a low profile and distance themselves from the dam decision' (Interview, June 1995, RSPB official). Under Abacha's regime, the National Institute for Policy and Strategic Studies was no longer so influential as it had been under that of Babangida.

Nevertheless, over the period of the judicial inquiry, NCF and HNWCP continued their advocacy efforts and made both formal and informal representations to the inquiry panel about the issues in the wetlands, giving the panel access to the wetlands project's scientific information. An NCF council member described one of his own informal advocacy efforts: 'The retired chief justice of Nigeria is a close friend of mine so I gave him the book [Hollis et al. 1993] to show the panel' (Interview, December 1994). This book, *The Hadejia-Nguru Wetlands: Environment, Economy and Sustainable Development of a Sahelian Floodplain Wetland* (Hollis et al.

1993), provided details of the case for conserving the wetlands, including the results of the hydrological model of the floodplain and the policy scenarios. By these means, NCF helped to ensure that the inquiry was supplemented by an investigation of the environmental impact of the dam project. In mid-1994 a national committee was set up, with its secretariat provided by the Federal Environmental Protection Agency (FEPA), 'to ascertain the environmental impact and macro-economic impact of water resource development projects in the basin; also, to come up with a management strategy for the water resources of the basin' (Interview, June 1995, wetlands project's new director, ex-FEPA). It was chaired by Dr Bukar Shaib, who had chaired the wetlands project's policy workshop in 1993, and its membership included a representative from the wetlands project.

Despite NCF's and HNWCP's lobbying, the report of the judicial commission of inquiry concluded that, by the year 2000, Nigeria would not be able to feed its population from rain-fed agriculture alone, and recommended that the government should proceed with the dam project. On receiving the report in May 1995, General Abacha reaffirmed his administration's determination to revitalise dam construction and irrigation development to enhance food production. NCF immediately organised a press conference to try to pre-empt the government's more fully considered response to the inquiry. Usually government responds to a judicial inquiry by means of a policy white paper, although in this case no such paper had appeared by the time Abacha's regime ended.

General Abacha died suddenly in June 1998. He was replaced by General Abubakar, who oversaw a peaceful changeover to a democratically-elected government under General Obasanjo in May 1999. Plans for the construction of the Kafin Zaki dam appear to have been quietly shelved. However, there are likely to be powerful interests working behind the scenes to ensure that dam construction is not shelved indefinitely. The return to democracy may mean that international agencies are more prepared to provide funding. But the longer that dam construction is postponed the more international opinion and environmental regulation will harden against such large-scale dam projects and towards more sustainable development. Meanwhile, the Hadejia-Nguru Wetlands Conservation Project has gained the right to be consulted on water distribution. They have encouraged links between upstream and downstream users, the state governors and the two river basin development authorities, thus improving awareness of the need for water sharing.

General lessons for NGO advocacy

Of significant importance in the wetlands campaign were the scientific expertise of the staff and researchers associated with the wetlands project and the political contacts of both NCF and the project. Writing on the influence of NGOs on environmental politics at the international level, Princen and colleagues have suggested that NGOs are most likely to make a distinctive and effective contribution when they can identify the political implications of bio-physical trends, that is, when they can make the link between the political and the scientific (Princen et al. 1994). The campaign to protect the wetlands provides an excellent illustration in the way that it made use of hydrological and economic data to demonstrate the impact of various policy options.

Princen and colleagues have suggested that another distinctive contribution that NGOs can make is to link the local with the global, to translate their experience at grassroots level into the political choices that have to be made at higher decision-making levels (Princen et al. 1994). In this respect, too, the campaign to protect the wetlands provides an excellent example.

The reason that a combination of the political and the scientific is so effective is that decision makers do not always have access to the type of detailed scientific knowledge that NGOs can gain when they undertake research to support an issue-specific campaign. They tend to rely on their scientific advisers. Conversely, scientists tend to focus on their own disciplinary interests and treat the political aspects with suspicion. Consequently, NGOs that can make the connections have a comparative advantage. Although Princen et al. were referring mainly to NGOs' contribution at the international level, the ideas also have relevance at the national level. In Nigeria, the military regimes have had little knowledge of environmental matters. As Areola notes:

> Lacking their own programmes for conservation and environmental management, successive Nigerian governments have come to depend very much on the ideas, concepts and programmes suggested or recommended by consultancy units, international agencies and seminars, workshops and conferences organised by local research and professional organisations.
>
> (Areola 1987: 287)

At the local level, government institutions such as the river basin development authorities have struggled to maintain basic functions, let alone gather scientific data. Such data as do exist are dispersed and patchy. So the wetlands project's data, particularly when translated by means of the computer model into the impacts of various policy options, represented a particularly powerful resource. NCF's political connections allowed that resource to be exploited to maximum effect.

The grassroots activity of the wetlands project gave NCF a legitimacy it might otherwise have lacked. One of NCF's weaknesses is that it can be viewed as an elitist organisation, with an urban membership drawn almost exclusively from Lagos in the south (although with eminent northerners among its council members), and a preoccupation with wildlife rather than people's more immediate concerns (although it is placing increasing emphasis on sustainable rural development, in line with WWF's changing conservation philosophy). In the case of the wetlands campaign, this perception of an elitist organisation was countered by the grassroots work of the wetlands project. Even though the initial focus of the project was on bird conservation, this was soon broadened to include a concern for the communities who depended on the wetlands for their livelihoods. In addition, HNWCP and NCF were able to argue that the interests of the downstream communities corresponded with the national interest by showing that maintaining agricultural production in the wetlands was likely to be more cost effective than expanding irrigated agriculture. This meant that the opposition of local communities and local governments in the downstream states to dam construction could be represented as supporting rather than challenging the federal government.

In Chapter 1, four contextual factors were mentioned as affecting an NGO's capacity to exert influence. These were: the nature of the specific issue and the interests involved; the character of the organisations being targeted by the NGO; the stage of policy development; and the political context with respect to the degree of democratisation. The wetlands campaign can be used to examine and expand upon these factors.

To take first the case of the specific issue and the interests involved, for the wetlands campaign the issue might easily have been depicted in polarised terms as the protection of a wildlife site of international importance from unsustainable development. As mentioned in the introduction to this chapter, the issue appeared to be a difficult one to influence, since the proposed dam was part of a long-standing programme for improving food production, to which the government was strongly committed. However, the HNWCP framed the issue in terms of water resources and rural community survival and development rather than wildlife protection and was able to argue convincingly, using its scientific and economic analyses, that its proposals served the national interest better than further dam construction. When HNWCP and NCF supported their case by referring to the international dimension, they did not mention the potential impacts on migrating birdlife, but instead warned of the risk of conflict with neighbouring countries that depended on the same water resources – a security matter of greater salience to Nigeria's military governments. NCF considered, but decided against, running a high-profile international campaign. It considered applying pressure to Berger's German parent company (Bilfinger and Berger) by enlisting the support of the German parliament and the German green movement, but decided against it, possibly because it recognised that the decision-making power did not lie with Berger. More importantly, such international pressure might merely have antagonised the military government and increased its resolve to continue with the dam construction programme. There was some low-level contact with Berger Nigeria, but this was mainly as a means of gathering information from sympathetic individual employees. HNWCP and NCF focused their main advocacy efforts on key decision makers in government at state and federal level within Nigeria, deliberately framing their arguments in terms of providing strategy support.

With respect to the second contextual factor, the character of the organisation targeted, HNWCP and NCF concentrated their efforts on government rather than the construction company, even though the military regime appeared to be a difficult target, lacking accountability other than to the army officer corps. HNWCP's and NCF's aim was to persuade government rather than to confront or embarrass it. Hence confrontation was viewed as counter-productive in much the same way as enlisting the support of international politicians and pressure groups was.

With respect to the third factor, the stage of policy development, again HNWCP and NCF appear to have been operating in a particularly difficult context. Chapter 1 suggested that, if the policy process is viewed as occurring in three stages, of agenda setting, policy formulation and policy implementation, then NGOs might be least able to exert an influence at the policy formulation stage. Yet, in the wetlands case, HNWCP and NCF appeared to achieve some success in persuading government to re-assess a long-standing policy commitment and to consider other options. Apart from the fact that they supported their case with scientific and economic evidence, another likely contributory factor is that they did not demand a particular outcome but instead presented a number of policy options that left plenty of room for negotiation and compromise. HNWCP and NCF did not demand that dam construction be halted altogether, but suggested a number of ways of reducing the impact and taking account of water needs further downstream.

With respect to the fourth factor, that of the degree or form of democratisation, again HNWCP and NCF might be viewed as operating in a difficult context. Virtually throughout the period of the wetlands campaign, HNWCP and NCF were working under a military regime. The literature on NGOs tends to assume that they will exert more influence in more democratic contexts. Yet, as an NCF council member commented, in some ways a military regime can be an easier target, especially if it would like to be seen as environmentally aware but lacks access to environmental expertise and experience. Furthermore, the policy community in a military regime is smaller and easier to identify than the diffuse policy networks of more democratic societies. Provided that HNWCP and NCF could convince the most senior government advisers of the merits of a particular course of action, the military regime could simply press ahead, since it had little need for extensive consultation to legitimise its policy with a policy network. The disadvantage is that since policies were not publicly debated and agreed, there was no obligation to carry a decision through if the regime changed. In addition, an opportunity to raise public awareness of the issues was lost. Close contacts established under one regime might not be useful, and might even be detrimental, if the regime changed, as NCF found to its cost.

As well as contextual factors, Chapter 1 referred to four characteristics of NGOs themselves that could affect their influence on policy. These concern their legitimacy and capacity, the appropriateness of their strategies, the way in which they frame issues and their networking with other NGOs. These characteristics are now examined in turn for their relevance to this case study.

The question of legitimacy has already been considered. NCF gained a legitimacy it might not otherwise have had from the grassroots work of the HNWCP among the wetlands communities. NCF and HNWCP also gained legitimacy by virtue of the expertise developed by the project staff on the hydrology of the Hadejia-Jama'are river basin. In terms of capacity, HNWCP gained considerably from their association with IUCN and the financial backing of their other international supporters, as well as from the input from overseas researchers. This kind of international backing is invaluable in a country like Nigeria, especially in an area as remote from the capital as the wetlands, since lack of transport and lack of reliable means of communication are major barriers to achieving influence for most Nigerian NGOs.

In terms of their strategy, both HNWCP and NCF deliberately avoided confrontation, preferring to work in a way that was complementary, so as to gain access to government officials. HNWCP shared its hydrological expertise and model with middle-level technical staff in state government, so that these staff could see for themselves the outcome of various policy options and use these as evidence to support the wetlands case in discussions with their seniors. While this in itself was insufficient to bring about a change in policy, it meant that officials were aware of other policy options and their likely consequences whenever political upheavals opened up opportunities for changes. NCF complemented this long-term strategy by publicising the HNWCP's findings at federal government level. Consciousness-raising activity by HNWCP among the downstream communities was fairly low key since it was demanding in terms of resources. NCF, although it supported general environmental education in schools, did not consider awareness-raising among the general public about the wetlands to be a priority. It was not an issue that could easily be presented in a way to rally widespread public support in Lagos, and in any case the government would be unlikely to heed urban public opinion on such a matter.

In line with their complementary rather than confrontational approach, HNWCP and NCF were careful to frame the issue in terms of the equitable sharing of water resources, a worthy aim that would be difficult to criticise. The issue might equally well have been defined as an issue about conserving migrating birds, or in terms of an outright opposition to all large-scale dams and irrigation schemes, but these would have been divisive and would have provoked strong counter-arguments from groups who saw their interests as threatened. The HNWCP and NCF recognised that outright opposition to the Kafin Zaki dam would be counter-productive, given the powerful interests in favour of the scheme. Instead they concentrated on highlighting the weaknesses of the existing and planned schemes and proposing alternative policy options for discussion.

As regards networking among NGOs, HNWCP and NCF for the most part acted independently but were able to provide backup for each other at critical times. Although there were tensions between the two organisations because of their different environments and styles of operation, they managed to maintain a common purpose and a united front on the wetlands issue. They avoided overt association with more radical groups who might 'rock the boat' by upsetting carefully cultivated relationships with government, although individual NCF officials had links with groups who favoured a confrontational approach and the adoption of a non-negotiable stance against the dam. Networking with international NGOs was limited to their funding organisations. NCF considered enlisting the support of other international environmental groups but only as a last resort.

To summarise, the campaign to protect the Hadejia-Nguru wetlands from the revival of an existing programme of dam construction represented a difficult challenge for HNWCP and NCF in several important respects. Yet they managed to achieve a number of concessions and contributed to the eventual suspension of further dam construction. In large measure the achievements were due to the complementarity between the wetlands project in northern Nigeria and the NCF in the south. This allowed powerful links to be made between their scientific and political expertise, and gave grassroots legitimacy to their high-level advocacy. HNWCP and NCF took care to ensure their arguments were aligned with local and national interests.

Some aspects of NCF's approach are distinctly Nigerian and may not be easily reproduced by NGOs elsewhere. In particular, NCF had access to government officials and the judiciary at the highest levels. In future, under a democratically-elected government, a more environmentally aware electorate may challenge NCF's claim to be acting in the best interests of the people, especially given its narrow membership base and lack of a democratic structure. However, from past evidence of NCF's political acumen, it is more likely that NCF will carefully adapt its approach and structure to any future changes in the political context.

Note on sources

Information about the wetlands campaign came from interviews with committee members and staff of NCF and the HNWCP, Ken Smith (RSPB), Henry Thompson (formerly project officer of the HNWCP) and officials in the Nigerian Federal Ministry of Water Resources and Rural Development.

Additional information came from the HNWCP Annual Progress Reports and publicity brochures, as well as the references listed below. Helpful comments on the chapter were provided by Tasso Leventis (NCF), Henry Thompson, Julian Thompson (University College London) and Mike Acreman (formerly IUCN).

References

Adams, W.M. and Hollis, G.E. (1988) *Hydrology and Sustainable Resource Development of a Sahelian Floodplain Wetland*, Report for Hadejia-Nguru Wetland Conservation Project to RSPB, Sandy, England and IUCN, Gland, Switzerland.

Areola, O. (1987) 'The political reality of conservation in Nigeria', in D. Anderson and R. Grove (eds) *Conservation in Africa: People, Policies and Practice*, Cambridge: Cambridge University Press.

Babadoko, S. (1995) 'FG to step up work on dams', *New Nigerian,* 31 May.

Barbier, E.B. and Thompson, J.R. (1998) 'The value of water: floodplain versus large-scale irrigation benefits in Northern Nigeria', *Ambio*, 27, 6: 434–40.

Barbier, E.B., Adams, W.M. and Kimmage, K. (1993) 'An economic evaluation of wetland benefits', in G.E. Hollis, W.M. Adams and M. Aminu-Kano (eds) *The Hadejia-Nguru Wetlands: Environment, Economy and Sustainable Development of a Sahelian Floodplain Wetland*, Gland, Switzerland: IUCN.

HNWCP (1993) *Hadejia-Nguru Wetlands Conservation Project Information Leaflet* No. 1, Nigeria: HNWCP.

HNWCP and NIPSS (1993) *Workshop on the Management of the Water Resources of the Komadugu-Yobe Basin, Kuru*, Nigeria: National Institute Press.

Hollis, G.E., Adams, W.M. and Aminu-Kano, M. (1993) *The Hadejia-Nguru Wetlands: Environment, Economy and Sustainable Development of a Sahelian Floodplain Wetland*, Gland, Switzerland: IUCN.

Hollis, Ted and Bedding, James (1994) 'Can we stop the wetlands drying up?', *New Scientist*, 2 July: 30–5.

NCF (1993) 'NCF delegation visits Shonekan', *Newsletter* (September), 2, 1: 1–5.

Olofin, E.A. (1993) 'Sustainable irrigation option for Nigeria: the large and the small', in E.M. Nwa and P. Pradhan (eds) *Irrigation Research Priorities for Nigeria*, Kano, Nigeria: International Irrigation Management Institute.

Potter, D. (1996) 'Democratisation and the environment: NGOs and deforestation policies in India (Karnataka) and Indonesia (North Sumatra)', in D. Potter (ed.) *NGOs and Environmental Policies: Asia and Africa*, London: Frank Cass.

Princen, T., Finger, M. and Manno, J.P. (1994) 'Translational linkages', in T. Princen and M. Finger (eds) *Environmental NGOs in World Politics*, New York: Routledge.

Thompson, J.R. and Hollis, G.E. (1995) 'Hydrological modelling and the sustainable development of the Hadejia-Nguru wetlands, Nigeria', *Hydrological Sciences*, 40, 1: 97–116.

9 The Ogoni people's campaign over oil exploitation in the Niger Delta

Susan Carr, Oronto Douglas and Uche Onyeagucha

The Ogoni people's campaign against Shell over oil exploitation in the Niger Delta, and over the associated environmental damage and alleged human rights abuses, has attracted international attention and support. This chapter chronicles the key events during the campaign by the Movement for the Survival of the Ogoni People (MOSOP), with particular reference to the period between 1993 and 1995. It analyses the responses of Nigeria's military governments and Shell and examines the way in which massive support for MOSOP's campaign was mobilised. It considers those aspects of the campaign's context and strategy that affected its chance of success. It concludes with a comparison of MOSOP's campaign with that of the Nigerian Conservation Foundation and the Hadejia-Nguru Wetlands Conservation Project, discussed in the previous chapter, to explore further the question of how NGOs may influence target institutions.

Origins and progress of the campaign

The grievances of the Ogoni people, like those of other communities in the Niger Delta, are long-standing, dating back to the British colonial era. They centre on the environmental damage done to their region by oil exploration and exploitation, their rights to a fair share of the oil revenue, and their oppression as a minority ethnic group by Nigeria's ethnic majorities. In the early 1990s, these grievances gained new force, encouraged by national and international pressures concerning democracy, the international focus on environmental issues and the emergence of a charismatic Ogoni leader in Ken Saro-Wiwa.

The Ogoni people live in the south east of Rivers state, on agricultural land cleared from the lowland rainforest of the delta (Figure 9.1). Oil exploration began in the Ogoni area in 1958 and Nigeria's first oil refinery was constructed there in 1965 (Osaghae 1995). The area has five major oil fields with ninety-six wells linked to five flowstations (in Bomu, Korokora, Yorla, Bodo West and Ebubu). Oil production is critical to Nigeria's economy. Throughout the 1990s it continued to provide about 80 per cent of the federal government's income and more than 90 per cent of Nigeria's

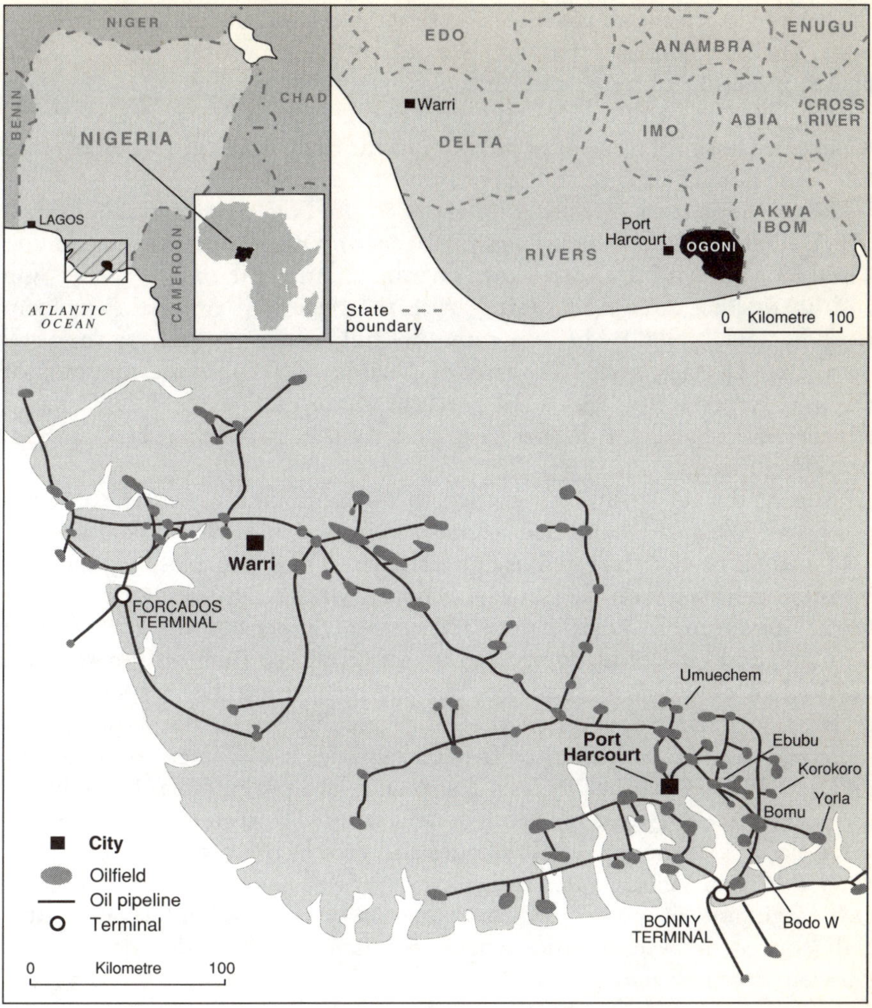

Figure 9.1 Area inhabited by the Ogoni people and distribution of oilfields in the Niger Delta.

foreign exchange earnings. In the early 1990s, the Ogoni area was producing about 1.5 per cent of Nigeria's oil, and had produced as much as 5 per cent in 1973 (Shell 1995).

There have been many serious oil spills, some of them dating back to attacks in the Biafran civil war between 1967 and 1970, when the Igbo in the east attempted to secede from Nigeria and the Ogoni were caught between the opposing sides. Oil spills continue to occur as a result of routine operations, the corrosion of ageing facilities and sabotage. The spills have polluted the land and water, affecting the livelihoods of the

Ogoni, who depend mainly on farming and fishing. Further pollution is caused by the flaring of gas after it is separated from the oil at the flowstations. The flaring means that villagers nearby live in constant light. In some places, pollution of fresh water has occurred when the cutting of channels for the installation of oil rigs and pipelines has led to the incursion of salt water. The huge increase in population in the area since oil was discovered has contributed to the overall environmental degradation.

The oil companies operating on-shore in the Niger Delta area are Shell (Anglo-Dutch), Agip (Italian) and Elf (French). Other companies, including Chevron and Mobil (American), operate mainly off-shore. Shell, Agip and Elf operate as a joint venture with the Nigerian National Petroleum Corporation (NNPC). Shell, or rather its fully-owned subsidiary the Shell Petroleum Development Company of Nigeria (SPDC), is the operator of the joint venture and has a 30 per cent share, whilst the NNPC has a 55 per cent share, Elf 10 per cent and Agip 5 per cent (SPDC 1995a; www.shellnigeria.com 1999).

Little of the oil wealth is returned to the communities where the oil is produced. When the Land Use Decree was passed in 1978, all land became held in trust by the state military government. Oil companies negotiate for oil mining leases with the state governors, who act as landlords for the federal government. Local people receive no compensation for the loss of farmland and little or no compensation for crops lost (Ashton-Jones 1996). In 1982, a government fund was set up, comprising 1.5 per cent of oil revenues, for development in the oil producing areas, but little of this money appears to have reached the communities it was intended to benefit (SPDC 1995b). Shell has its own community aid programme, but some of its projects are criticised for being poorly attuned to community needs and being badly maintained. Most villages still lack health facilities, schooling, accessible roads, clean water and electricity. Shell has publicly stated that it gave large sums of money to Ogoni community projects but 'Privately, officials conceded these donations never reached their intended destination' (Rowell 1995).

Community leaders in the Niger Delta, especially among the Ogonis, see their communities as the most politically marginalised in the country (Osaghae 1995). They accuse the majority ethnic groups (the Hausa-Fulani in the north, the Yoruba in the west, and the Igbo in the east) of using oil wealth to develop their areas at the expense of the areas where the oil is produced.

Although these grievances had been forcefully aired many times before, they were argued with renewed vigour and a heightened sense of injustice after a number of existing Ogoni organisations joined forces to form the Movement for the Survival of the Ogoni People (MOSOP) in 1990. The stimulus for the formation of MOSOP came from Ken Saro-Wiwa, an internationally-known writer with experience of government administration, business and academia. At the time he was president of the Ogoni

Central Union, the biggest Ogoni communal organisation, which was itself formed to unify disparate groups in 1945 so as to compete more successfully for power and resources.

A number of factors may have led Ken Saro-Wiwa and other Ogoni leaders to decide that the early 1990s was an opportune time to press once again for more power and resources for the Ogoni people. There was a general heightening of political activity, with steps being taken towards democratic elections and a lifting of the ban on political parties in 1989 (Lewis 1994). In mid-1990, there was a rapid increase in oil prices as a result of the Persian Gulf crisis, creating a mini-boom in Nigeria. There was increasing unrest among the oil communities, with more than sixty protests against Shell in 1990 (Rowell 1995).

Describing the formation of MOSOP, Ken Saro-Wiwa later recollected:

> I began to mobilize the Ogoni people to protect their environment, their culture and society with the publication in March 1990 of my diary on the Nigerian Civil War, *On a Darkling Plain*, one of the most controversial books ever published in Nigeria. ... I made wide consultations with chiefs, elders, women's groups, the elites and associations, with a view to forming a common ground to demand a stop to the mindless destruction of our environment, halt the continuing degradation and request appropriate compensation.
>
> (The Ogonis 1995, quoted in Douglas and Onyeagucha 1995)

As a focus for those consultations, Ken Saro-Wiwa drafted the Ogoni Bill of Rights. This summarised the grievances of the Ogoni people and made a number of demands. These included the right to political control of Ogoni affairs by Ogoni people, as well as the right to control and use a fair proportion of the Ogoni economic resources for local development, to protect the Ogoni environment from further degradation, and to be adequately and fairly represented in all Nigerian institutions. After minimal revision by Ogoni chiefs and leaders, the Bill of Rights was presented to General Babangida's military government in August 1990. Although its receipt was acknowledged, there was no other response at that time.

MOSOP then launched a vigorous campaign to publicise the Ogoni cause and to enlist support both locally and internationally. Towards the end of 1990, a demonstration by a neighbouring ethnic group against Shell's plant in Umuechem was brutally suppressed by Nigeria's notorious Mobile Police Force, leaving eighty people dead. This renewed the determination of Ogoni activists to pursue their cause. A new, more radical, Ogoni group was formed, the National Youth Council of Ogoni People (NYCOP), made up of graduates and professionals, who urged MOSOP to take further action to pursue its goals. In December 1992, after an extensive tour of Ogoni communities to enlist their support, MOSOP's leaders sent an ultimatum to Shell, Chevron and NNPC. They demanded payment of

US$ 6 billion for accumulated rent and royalties since 1958, US$ 4 billion as compensation for environmental damage, an immediate stop to environmental damage and gas flaring, the immediate covering of all overground high pressure pipelines, and the initiation of negotiations with the Ogoni people about the future conditions for further oil exploration and exploitation (Osaghae 1995). They threatened mass action to disrupt the oil companies' operations if these demands were not met within thirty days. In response, the oil companies tightened their security and the federal government sent in troops to protect oil production. The government banned all public gatherings and demonstrations.

On 4 January 1993, the day after the deadline expired, more than 300,000 Ogoni (of a total population estimated at 500,000) defied the government ban and took part in a protest march and mass rallies to repeat their demands and tell Shell to leave the area. Beforehand, the government attempted to get Ken Saro-Wiwa to call off the protest. He and other MOSOP leaders were arrested and detained over Christmas, but released before the protest took place. With the heightened tension making conditions dangerous, Shell withdrew its staff from the Ogoni area the next day, although automatic control of the flowlines allowed some oil production to continue until the middle of the year. Further mass demonstrations of support for the Ogoni cause followed. For example, on 13 March 1993 a vigil was held in churches throughout the Ogoni area to pray for the deliverance of the Ogoni people. The same month, an Ogoni Survival Fund was launched, to which each Ogoni contributed 1 Naira, the equivalent then of 5 pence.

That year the government issued a decree, the Treason and Treasonable Felony Decree, declaring demands for rights to self-determination and disturbances of oil production to be acts of treason punishable by death. On 12 June 1993, the day of the federal presidential election, most Ogoni people followed the advice of their leaders and refused to vote. Ken Saro-Wiwa was arrested on 22 May for 'election offences' and accused of 'sedition and unlawful assembly'. He was released on bail a month later (21 June) by which time the election had been held and annulled.

Intimidated by the possibility of arrest for treason, the president and vice-president of MOSOP resigned. In their place, Saro-Wiwa, whose role in MOSOP had previously been that of spokesperson, was elected president (although he was still in prison at the time) and an official of the Civil Liberties Organisation (CLO) was elected vice-president. With these new appointments, MOSOP underwent fundamental change (Douglas and Onyeagucha 1995). Drawing on the extensive experience of the CLO as Nigeria's foremost human rights group, MOSOP became more structured and organised. Offices were opened in Port Harcourt and Lagos, a formal constitution was approved, an Ogoni anthem was written and an Ogoni flag produced. Other organisations were accepted as affiliates of MOSOP (Figure 9.2), including NYCOP, the Conference of Ogoni Traditional

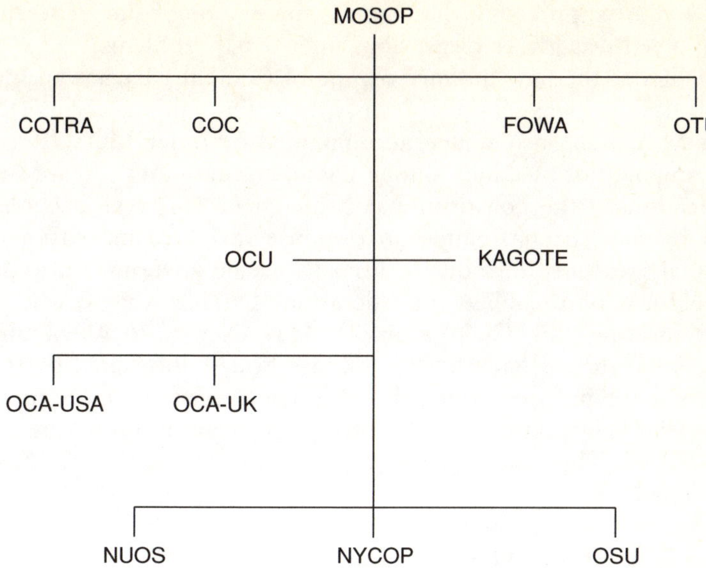

Figure 9.2 The Movement for the Survival of the Ogoni People and its affiliated groups.

Key
COC – Council of Ogoni Churches
COTRA – Conference of Ogoni Traditional Rulers
FOWA – Federation of Ogoni Womens' Associations
KAGOTE – umbrella organisation for Ogoni traditional rulers
MOSOP – Movement for the Survival of the Ogoni People
NUOS – National Union of Ogoni Students
NYCOP – National Youth Council of Ogoni People
OCA – Ogoni Community Association
OCU – Ogoni Central Union
OSU – Ogoni Students Union
OTU – Ogoni Teachers Union

Rulers (COTRA), the Council of Ogoni Churches (COC), the Ogoni Teachers Union (OTU), the Council of Ogoni Professionals (COP), the Federation of Ogoni Womens' Associations (FOWA), the National Union of Ogoni Students (NUOS) and the Ogoni Students Union (OSU). A branch of MOSOP was established in the USA and Ogoni people in the UK organised to lend their support to the cause. An offshoot of the Civil Liberties Organisation, Environmental Rights Action (ERA), was formed to campaign for the rights of people to a healthy environment.

Between July and September 1993 there were violent clashes between the Ogonis and neighbouring Andonis following apparently unprovoked attacks. Both sides suffered casualties. More than 1,000 Ogonis were killed and 30,000 made homeless. Similar clashes occurred between the Ogonis and the Okrikas in December. MOSOP leaders claim that these attacks were deliberately provoked by the military in collusion with the oil companies as a way of suppressing the Ogonis, since none of the leaders of the groups involved was aware of previous inter-tribal problems.

In 1994, under the new military regime of General Abacha, an Internal Security Task Force was set up to destroy dissent among the Ogoni (Rowell 1995). In May, Ken Saro-Wiwa, accompanied by other MOSOP leaders, began a campaign of meetings among Ogoni communities to gain support for his election to the constitutional conference. This was a conference convened by the Abacha regime, in response to increasing national and international pressures for a democratically-elected government, to discuss the future form of the Nigerian federation. On his way to one of his campaign meetings, at Giokoo on 21 May, Ken Saro-Wiwa and his entourage were turned back by the military police. Infuriated by this, an angry mob at the meeting attacked and killed four of the more moderate Ogoni leaders, believing them to be pro-government and working against Ken Saro-Wiwa. The next day, Ken Saro-Wiwa was once again arrested, accused of inciting the murders, even though he and MOSOP had always advocated non-violent protest.

At this point other MOSOP leaders and Ogoni activists went underground. FOWA and COTRA disbanded. COC, OTU, NUOS and OSU all adopted a low profile. However, a number of environmental groups in Nigeria with a special interest in protecting the Niger Delta continued to take up environmental issues with Shell, for example, ERA, the Niger Delta Wetlands Centre, Pro-Natura, and Niger Delta Watch. Both the human rights issues and the environmental issues, which were becoming increasingly interrelated, were pursued in a vigorous campaign against Shell at the international level. For example, in the UK the Body Shop (a cosmetics retail chain that prides itself on its ethical principles) took up the Ogoni cause after its founder and chief executive, Anita Roddick, had been inspired by Ken Saro-Wiwa at a meeting in the Netherlands on the plight of indigenous peoples organised by UNPO (Unrepresented Nations and Peoples Organisation). Body Shop mounted shop window displays about the plight of the Ogoni and Ken Saro-Wiwa, painted slogans of support on the lorries transporting its goods and took out full-page advertisements in the UK national press to urge consumers to boycott Shell petrol stations. Pen International, a writers' organisation that campaigns on behalf of writers who become political prisoners, took up Ken Saro-Wiwa's case with the Nigerian government. The Gaia Foundation provided support for the Ogoni cause after hearing a talk by ERA at a meeting convened by the UK's Royal Geographical Society on 'Petroleum and Nigeria's Environment'. Human Rights Watch in the UK and the USA also took up the cause of Ken Saro-Wiwa and the other Ogonis imprisoned with him.

Together the Body Shop, the Gaia Foundation, Greenpeace, Friends of the Earth, Amnesty International, Pen International, UNPO and Ogoni people living in the UK formed a loose coalition to raise public awareness of the issues. Individuals in the coalition took responsibility for pursuing different activities: lobbying parliament, securing media coverage, meeting with Shell and networking with other NGOs. They lobbied organisations such as pension funds with large shareholdings in Shell and encouraged demonstrations at shareholders meetings.

In February 1995, when the trial of Ken Saro-Wiwa began, more than 26,000 Ogonis mounted a demonstration at the premises of the special military tribunal, despite the threatening circumstances. On 31 October Ken Saro-Wiwa was found guilty and sentenced to death. Apparently impervious to national and international opinion, and pleas for clemency that included those from the British government and the Commonwealth, the Nigerian military regime went ahead with the sentence. On 10 November, Ken Saro-Wiwa and eight other Ogonis were hanged.

This sequence of events is summarised in Table 9.1. The day after the hangings, Commonwealth nations expressed their deep dismay by suspending Nigeria's membership of the Commonwealth. Amnesty International commented that, after the hangings, 'The shock waves reverberated around the world. For a moment, the international community looked set to direct the force of its disapproval powerfully and effectively against an unrepentantly brutal regime. Then the spotlight moves elsewhere' (Amnesty 1996). Temporarily at least, Abacha's military government managed to suppress the Ogoni issue. However, the main issues remain unresolved. As of September 1999, Shell has still not returned to the Ogoni area and has no plans to return until invited back by the community. The simmering sense of injustice among communities in the oil-producing areas remains. The UK's *Guardian* newspaper reported that as Nigeria's democratically elected President Obasanjo took over from General Abubakar's military regime he faced renewed unrest: 'as Obasanjo takes over he faces unrest in the oil-rich Niger Delta that threatens the country's financial lifeline. Delta communities are increasingly strident in demanding a share of the riches. Sabotage and kidnappings interspersed with violence have shaken the oil industry, though so far outright fighting has been sporadic' (McGreal 1999). The unrest is now spreading to all minority ethnic groups in the delta (Vidal 1999).

What was achieved?

MOSOP deliberately adopted a high-risk strategy: direct confrontation of a military regime over an issue of critical importance to the national economy, namely oil production. They framed their campaign in terms of high moral principles, such as human rights, justice and truth, against a government for whom these concepts appeared to have little meaning. When the military regime responded at first with indifference, then with increasing repression and force, MOSOP sought to bring pressure to bear on the government indirectly, by targeting Shell and by seeking support

Table 9.1 Timing of key events (action by protesters and reaction by government and SPDC) in the Ogoni people's campaign over oil exploitation in the Niger Delta

	Action by protesters	Reaction by government/SPDC
1990	More than 60 protests against Shell in the Niger Delta	
	MOSOP formed; MOSOP and traditional leaders present Ogoni Bill of Rights to General Babangida's military government, demanding political autonomy (26 August)	
1991	Press conferences to explain Ogoni action and elicit support of pro-democracy groups	
1992	MOSOP leaders write to Shell, Chevron and NNPC demanding rent and royalties, compensation, an end to environmental degradation, and negotiations about future conditions for oil exploitation, giving the companies 30 days to respond (from 3 December)	OMPADEC set up to co-ordinate development projects in oil-producing areas SPDC tightens security and federal government send in troops; government bans all public gatherings and demonstrations
	NYCOP formed	Ken Saro-Wiwa and other MOSOP leaders detained over Christmas ahead of January rally
1993	Mass protests and rallies, supported by 300,000 people, to press Ogoni demands (4 January)	SPDC withdraws its staff from Ogoni area in response to intimidation and threats from communities
	Mass vigil and prayers in churches throughout Ogoniland (13 March)	Government issues Treason and Treasonable Felony Decree
	Ogoni Survival Fund launched (March) In response to a call from MOSOP, many Ogonis refuse to vote in federal election for president (12 June)	Ken Saro-Wiwa arrested for 'election offences' relating to the 12 June election (which is subsequently annulled); released on bail a month later
	MOSOP's president and vice-president resign, intimidated by threats of prison; Ken Saro-Wiwa elected president, while still in prison, and Ledum Mittee (of CLO) as vice-president	Violent clashes between Ogonis and other clans, following apparently unprovoked attacks (July-December; MOSOP believe the attacks to be instigated and financed by SPDC with the support of government)
1994		Abacha's military regime sets up Internal Security Task Force, to destroy dissent among oil communities (January)
	A series of meetings held to mark Ogoni Day (21 May); four moderate Ogoni leaders attacked and killed	Ken Saro-Wiwa detained by military on his way to address one of the Ogoni Day meetings
	Other MOSOP leaders go underground	Ken Saro-Wiwa arrested on 22 May, accused of inciting the murders

cont./

Table 9.1 cont

1995	More than 26,000 Ogoni people protest outside the tribunal premises	Ken Saro-Wiwa's trial by special tribunal begins (February)
	International action (e.g. in UK: demonstrations outside Nigerian High Commission, protests at annual general meeting of Shell shareholders in UK, Body Shop and Gaia Foundation co-ordinate campaign with Greenpeace, Friends of the Earth, Pen International and Ogonis living in the UK)	Niger Delta Environmental Survey launched Ken Saro-Wiwa found guilty and sentenced to death (31 October) Ken Saro-Wiwa and eight other Ogonis hanged (10 November), despite pleas for clemency from British government, the Commonwealth and Amnesty International

Sources
Osaghae (1995), Rowell (1995), Douglas and Onyeagucha (1995).

Abbreviations
CLO – Civil Liberties Organisation
MOSOP – Movement for the Survival of the Ogoni People
NNPC – Nigerian National Petroleum Company
NYCOP – National Youth Council of Ogoni People
OMPADEC – Oil Mineral Producing Area Development Commission
SPDC – Shell Petroleum Development Corporation of Nigeria

from the international community. This seemed only to make the government more determined to suppress the Ogoni communities and silence the people it saw as the leaders. Many Ogoni people paid with their lives. So a question that needs to be asked is: did MOSOP's campaign lead to any lasting achievements, or were those lives lost in vain? Were there any positive responses either from government or Shell?

One outcome was that government increased the resources allocated for community development in the Niger Delta. In 1992, there was an inquiry into the distribution of funds to the oil producing areas (SPDC 1995b). Subsequently the funds were doubled from 1.5 per cent to 3.0 per cent of oil revenues and a new organisation, the Oil Mineral Producing Area Development Commission (OMPADEC), was established to manage them. In 1995, government announced that it would increase these funds to 13 per cent (Shell 1995). Although that pledge was never fulfilled, and little of the money channelled through OMPADEC reached the communities, there is more recognition that they have been treated unfairly. Obasanjo's democratically-elected government has been urged by senate to ensure that 3 per cent of oil revenues is used for the communities' immediate benefit.

Another outcome was that there was some small recognition of the human rights issues by the military regime that succeeded General Abacha. In 1998, General Abubakar said that he would not allow any repeat of the human rights abuses committed by the previous regime (Smith 1998). He said that Ogoni activists arrested since the execution of Ken Saro-Wiwa would be tried by civil courts, not by a special tribunal.

Shell's initial response to the campaign was mainly defensive, although under continuing pressure from its European shareholders it eventually became more conciliatory. From the start, the company expressed some sympathy for the view that not enough of the revenue from oil was going back to the inhabitants of the oil-producing areas. In a briefing document Shell stated: 'As a measure of our concern for these communities, we have made this view known whenever appropriate, and this support has undoubtedly helped in the raising of the proportion of revenues going to oil producing areas' (Shell 1995).

However, Shell claimed that most of MOSOP's demands were outside the business scope of oil operating companies and within the government's sphere of responsibilities (Shell 1995; SPDC 1995b). When Shell came under pressure to intervene on Ken Saro-Wiwa's behalf after the special tribunal's decision, it issued a statement calling for 'quiet diplomacy', adding: 'We believe that to interfere in the processes, either political or legal, here in Nigeria would be wrong. A large multinational company such as Shell cannot and must not interfere with the affairs of any sovereign state' (Rowell 1995). Critics such as Body Shop and Greenpeace saw this policy as tantamount to backing the military regime. Shell has since come under continuing pressure from some of its shareholders to revise its stated aim of being apolitical. A report by Pensions & Investment Research Consultants Ltd for its clients, namely pensions and investment funds with actual or potential shareholdings in Shell, noted that 'given the size of Shell's interests in Nigeria, its joint venture with government and the central place it occupies with the country's economy, Shell cannot in practice adhere to a policy of non-interference' (PIRC 1996). It recommended that Shell should develop a statement of business ethics to guide its operations in countries where there are abuses of human rights. Shell subsequently admitted shortcomings in these aspects of its policies. It has since incorporated the language of human rights into its operating principles and has consulted with human rights organisations (Ganesan 1998).

In response to criticisms of its environmental record in the Niger Delta, Shell argued that over-fishing, deforestation, poor agricultural practices and over-population were more significant environmental problems than oil pollution and gas flaring. Whilst still holding to this view, in 1995 Shell initiated a long-term survey, the Niger Delta Environmental Survey, to provide baseline ecological and socio-economic data. An independent steering committee, with one representative from Shell, drew up the terms of reference for the survey and appointed the consultants. Funding was provided at first by Shell and then by an umbrella group representing oil producers operating in Nigeria. Phase 1 of the survey, reviewing existing information, was completed in 1997. Phase 2, conducting field surveys, was expected to be completed early in 2000. The results are intended to serve as a guide for all future developments in the area, not just oil production.

Regarding oil spills, Shell argued in its defence that 28 per cent of oil spills between 1989 and 1994 were due to sabotage, while admitting that others were due to routine operations and the corrosion of ageing facilities (SPDC 1995a). Shell announced that it was continuing its programme of replacing all flowlines older than fifteen years situated in mangrove swamp, and renewing land flowlines where necessary. It was continuing to upgrade its flowstations. It was providing training for its staff to increase awareness of their responsibilities towards the environment and local communities.

In response to the criticisms about gas flaring, Shell argued that gas is an unavoidable by-product of oil production and that flaring is necessary because there is only a small local market for gas. It said that originally the gas flares were sited away from communities but that as the population has increased communities have encroached on the land nearby. It is attempting to reduce the problem in three ways. The first is by injecting the gas back into the oil reservoirs, although this is only possible in a few sites. The second is by encouraging local markets for the gas, although this is difficult while domestic oil prices are low. The third is by developing export markets, for example in Europe. For this purpose it has a 24 per cent stake in a project to develop a liquefied natural gas plant in the Niger Delta. However, this is a complex, expensive and long-term scheme that is not expected to generate a profit until 2007 (SPDC 1995a). Gas that cannot be removed by these means will still have to be flared.

On environmental matters in general, critics have accused Shell of operating double standards. Shell has admitted that it tailors the implementation of its environmental policy to meet local circumstances, in that 'individual Shell companies set their own detailed goals and targets, reflecting the national, legal and cultural background in which they work' (SPDC 1995a). A Shell environmental adviser said: 'Shell here is Nigerian: Shell Petroleum Development Corporation, SPDC. There is no environmental awareness', adding 'In Nigeria Shell has old equipment, it's a minority shareholder – these do not make the situation easy from an environmental point of view' (Interview, June 1995). Commenting on Shell's efforts to improve its environmental performance, Pensions & Investment Research Consultants noted, with respect to environmental protests by local communities and outsiders, 'there are question marks over whether, without these pressures, application of Shell's current environmental policy would have resulted in the commitments to environmental improvement now evident' (PIRC 1996). The report concluded 'international companies have clear responsibilities to operate to the highest standards and to be acceptable to local communities wherever they operate'. It recommended that Shell should review its environmental policy in this respect.

In summary, it seems that both government and Shell were prompted to reconsider and revise some of their policies in response to the Ogoni campaign, even though they might deny this was a cause. Although the response fell far short of the original demands, and was achieved at high cost, there were a number of positive outcomes. These include the establishment of OMPADEC and increased funding for community development, the Niger Delta Environmental Survey, and more attention to ameliorating the environmental impacts of oil production. In addition, there is now an international awareness of the human rights and environmental issues in the Niger Delta so that the actions of Shell in Nigeria remain under scrutiny.

How MOSOP gained support for its cause

One of the outstanding achievements of the MOSOP campaign was the tremendous support and publicity that it attracted, both locally and internationally. This section examines how that support was gained, drawing extensively on the first-hand account provided by Douglas and Onyeagucha (1995).

Communication, by means of extensive networking and use of the media, was central to the campaign. The speed with which messages were transmitted to the Ogoni communities, from the communities to the MOSOP leaders, and outwards to the media and supportive groups abroad, is impressive given the general difficulty with telecommunication links in Nigeria and the particularly poor infrastructure in the Niger Delta.

As already mentioned, local networking began with consultations to enlist the support of the communities' traditional chiefs and leaders, and with public meetings to discuss the issues with local community-based associations. As a result, MOSOP was formed from the amalgamation of several existing groups and organisations, including the Ogoni Central Union, the Kagote (an umbrella group of Ogoni elites and traditional rulers) and the National Union of Ogoni Students (an association of Ogoni students in higher education). Recognising the importance of religion to the communities (95 per cent of the Ogoni population are Christian), the MOSOP leadership encouraged the churches to band together to form the Council of Ogoni Churches. Similarly, it encouraged traditional leaders to unite as the Council of Ogoni Traditional Rulers. Later, these and many other existing Ogoni organisations officially became affiliate members of MOSOP (Figure 9.2).

All MOSOP's and NYCOP's activists were encouraged to attend a course of 'field orientation' lectures given by Ogoni leaders and invited experts. Topics included history, economics, political science and sociology in relation to the Niger Delta and the Ogoni people. The aim was to strengthen their resolve to pursue the Ogoni cause whatever the odds against them.

Powerful symbolism was used to unite the Ogoni communities behind a common cause. The Ogoni Bill of Rights built on the people's everyday experience of the impact of oil pollution and their sense of injustice at their hardship amidst the evidence of oil wealth. As Ken Saro-Wiwa noted about the public speaking tours by MOSOP leaders to raise awareness among the communities:

> I found that I was not telling the Ogoni people anything that they had not known or were not expecting. Children, men and women had lived their determination patiently and were only waiting for a spark to ignite the fires that glowed in them. All the steering committee had to do was to provide the leadership.
>
> (The Ogonis 1995)

Later, the Ogoni anthem and flag served as further unifying symbols and constant reminders of the injustices suffered.

The military government, ethnic majority groups and particularly Shell were depicted as 'the enemies'. Among the oil companies, Shell was singled out partly because it is a high profile company, not necessarily because it was the worst offender: 'Chevron may be the worst, or next to Agip' (Niger Delta ecological expert interview, June 1995). This motive was recognised by Shell: 'There are other oil companies in the area, but we're about as big as all the others combined. It's more glamorous to attack us', and 'For hard core Shell people there's a sense of injustice that Shell is picked on' (Shell environmental adviser interview, June 1995). Shell was thus used as a symbol, to represent the oppressor or villain in MOSOP's campaign.

By means of the network of groups, the whole of Ogoni society had links to MOSOP through one organisation or another and could be readily mobilised to take action. For example, most big events, such as the mass rally on 4 January 1993 after the ultimatum to Shell expired, were preceded by church services at which the details were announced and prayers said for the survival of the Ogoni nation. In addition, NYCOP toured all the main Ogoni regions and set up committees in each village to mobilise support for the mass rally. As well as church services, there were marches, rallies, speeches, dance, drama and cultural displays in the communities the day before the mass rally.

NYCOP coordinators in the villages and towns made use of local groups (such as market women's groups) to disseminate information. The coordinators had to ensure that activities and events were reported back to MOSOP headquarters. MOSOP printed a newsletter, the *Ogoni News*, which was circulated on market days in the main town of Bori and sent to NYCOP coordinators for further distribution. Traditional forms of communication such as 'town criers' helped spread news items. Impending activities were also announced by means of paid advertisements on government-controlled radio, until the murder of the four Ogoni leaders when this form of dissemination was banned by the government.

All possible means were used to publicise the Ogoni cause more widely. Writers and journalists in MOSOP used the state and national press to comment promptly on any events in Nigeria that could be linked with the Ogoni issues, wrote deliberately provocative articles to stimulate further discussion and responded to critical articles. Ken Saro-Wiwa referred to the

Ogoni cause in his regular column in the Nigerian *Sunday Times* and in his international speaking engagements. MOSOP issued press releases and briefed correspondents. All MOSOP's press releases, briefing notes and articles were sent by fax to supportive human rights and environmental groups in Nigeria and the UK and later in the USA. In the UK, the Gaia Foundation helped with onward transmission to its own mailing list of supporters. MOSOP sent speakers to the UK and the USA to brief the Ogoni expatriate community as well as journalists and supportive organisations. Journalists and groups such as Amnesty International, Greenpeace and Action Aid were invited to the Niger Delta to see the situation for themselves. MOSOP activists in the communities helped provide video footage for two documentary programmes on the issues: *The Heat of the Moment* and *The Drilling Fields*, which were broadcast in the UK and widely shown among pressure groups in the UK and USA. MOSOP leaders presented the Ogoni Bill of Rights to international bodies with a human rights remit, such as the United Nations subcommittee of Human Rights on the Prevention of Discrimination Against and Protection of Minorities, and the General Assembly of the Unrepresented Nations and Peoples Organisation.

When MOSOP was forced to go underground in Nigeria, networking by supportive groups outside the country took on its own momentum. For example, in the UK, campaigners including road protesters, human rights activists and Greenpeace joined under the title 'Shell Out' to protest outside Shell's annual general meetings from May 1995 onwards. Body Shop held regular Ogoni campaign meetings with UK Ogoni associations and organisations such as Greenpeace and Amnesty International, at which they discussed how to lobby Shell's major shareholders. These activities overlapped with the campaign against Shell in Europe in mid-1995 over its plans to dispose of its redundant Brent Spar oil rig in the North Sea. The resulting extensive adverse publicity meant that Shell was forced to take both campaigns very seriously.

Lessons for NGO tactics and strategies

As an organisation, MOSOP cannot readily be categorised. It is neither purely an environmental NGO nor purely a political liberation movement but a cross between the two, of the kind described by Mohamed Salih (1999: 60). Such hybrid movements, Salih argues, are symptomatic of situations where the environment and livelihoods of rural communities have been put at risk by authoritarian governments. Deprived communities, embittered by their treatment at the hands of the people who hold all the power and excluded from any democratic say in decisions of critical importance to them, resort to resistance and protest.

The Ogoni campaign, led by MOSOP, used a high-risk strategy of confrontation against a military government. One of the main reasons that it made any headway at all was because it complemented this strategy with consciousness-raising, gaining tremendous support from the Ogoni people and from the international community. The Ogoni issue became known internationally as a symbol for human rights causes. This section reflects on the lessons for NGOs more generally, on the basis of the framework set out in Chapter 1. It considers first the contextual factors that affected the campaign and then the characteristics of MOSOP. It concludes by comparing these factors and characteristics with those of the Nigerian Conservation Foundation's and the Hadejia-Nguru Wetlands Conservation Project's campaign to protect a wetlands area, described in the previous chapter.

To take first the nature of the issue, MOSOP's demands on behalf of the Ogoni people represented an extreme challenge to the Nigerian government. The revenues from oil production were fundamental to Nigeria's economy, so claims to a much greater share of the oil wealth by a minority group, and threats to disrupt oil production if these claims were ignored, would inevitably be fiercely resisted. Demands for self-determination similarly represented an extreme challenge to government, threatening to loosen federal ties and even raising the spectre of the secession of oil-producing states. Demands for environmental improvements, while they might be resisted by government and oil companies on the grounds of cost, represented a lesser threat. However, they might not have unified the Ogoni community behind the cause to the same extent as the other issues. Taking the nature of the issue alone, one might predict that the likelihood of NGO influence would be virtually zero, except possibly in relation to the environmental aspects.

Turning to the nature of the target organisations, MOSOP's principal targets were the federal government and Shell. Almost throughout the main period of the MOSOP campaign (1990–95), Nigeria's government was a military one, led first by General Babangida (1985–93) then, after a military coup, by General Abacha (1993–98). There was a brief three-month period in 1993 when control was handed to an interim civilian government under Chief Shonekan. The military government, while for the most part tolerating outspoken criticism, for example in the lively and widely read Nigerian press, responded with force whenever its interests were seriously threatened. Periodically it closed down certain newspapers and banned public meetings when it viewed these as disruptive and divisive. So, given the threats to economic and political interests posed by the Ogoni people's demands, MOSOP stood little chance of influencing the government directly. Shell was a softer target. As an international company, it prided itself on its reputation. To be accused of ignoring human rights and of applying weaker environmental standards in Nigeria than in northern countries was damaging to its image. With its headquarters in the Netherlands and the UK, countries with strong environmental pressure groups, it was vulnerable to environmental campaigns and consumer pressure in

Europe. Within Nigeria, it was viewed as having close and influential links with government. Shell was therefore seen by MOSOP and its supporters as offering a point of leverage on government.

To examine next the stage of policy development, on the basis of a 'three-stage' view of the policy process MOSOP's campaign can most appropriately be described as agenda-setting, rather than providing policy choices or contributing to policy implementation. It challenged the status quo and demanded changes to existing policies rather than offering policy options. NGOs are generally viewed as having most opportunity for influence at the stage of policy implementation, and least at the stage of decisions among policy choices (see Chapter 1). They can sometimes be influential in agenda-setting, especially if they provide convincing evidence of the need for a new agenda, or if policy makers are searching for new ways to solve complex and intractable problems. However, MOSOP's demands, which included the right to self-determination, to control and use a proportion of the revenue from oil and to receive back-payments, represented a fundamental challenge. Such demands would be unlikely to be received sympathetically by any government, let alone a military one.

So the context in which MOSOP's campaign was fought was an extremely unpromising one, apparently offering little political space for NGO influence, except for the potential vulnerability of Shell in relation to the environmental issues.

The characteristics of the NGO are now examined, to explain how MOSOP achieved some successes despite the unpromising context. MOSOP's campaign gained strength from the legitimacy conferred by its massive grassroots support. This support was demonstrated by the huge numbers of Ogoni people that turned out at MOSOP's rallies to demonstrate their solidarity with the cause, despite the threat of retaliation. MOSOP's capacity was enhanced by Ken Saro-Wiwa's international status and connections, and more generally by the experience and ability of its leaders. Wide consultation with influential people and groups in the communities, impassioned speeches among the communities, and material help and comfort for families harmed during the campaign, made Ken Saro-Wiwa a venerated figurehead for most Ogoni people.

MOSOP's confrontational strategy was therefore complemented by its strategy of consciousness-raising. It worked hard at making the Ogoni people fully aware of all the issues, building on their own experience and sense of injustice. It kept the communities informed of events by means of community leaders, coordinators, newsletters and 'town criers'. It rallied them to the cause with speeches, prayers, song and drama. It united the communities around symbols such as the Ogoni Bill of Rights, and the Ogoni flag and anthem. MOSOP made the wider national and international community aware of its cause with newspaper articles, briefing papers, speaking engagements and video material, and by encouraging visits from journalists and campaigners.

In line with its dual strategy of confrontation and consciousness-raising, MOSOP framed the issue in terms of fundamental principles, basing the campaign on human rights and justice. It spelled out its objectives in uncompromising terms, with ultimatums. It depicted the government, Shell and ethnic majorities as 'enemies'. In these ways, in confronting its target groups it further unified its supporters.

MOSOP strengthened its hand against government and Shell by its extensive networking. By linking together church leaders and by linking up with other Ogoni groups such as the Ogoni Teachers' Union and the Federation of Ogoni Women's Associations, it had the capacity to reach every member of the Ogoni communities. By linking with the national Civil Liberties Organisation through MOSOP's vice-president, it was able to draw on that organisation's experience and contacts. MOSOP's international links with human rights and environmental groups enabled it to carry on its campaign even when it was forced to go underground. International networking can be seen as a threat to national sovereignty, so provoking a backlash from government (Eccleston 1996). In this case, repression by government was already so severe that this risk was irrelevant. MOSOP hoped that pressure on the Nigerian government from the international community would at least protect the lives of the Ogoni people on trial.

A comparison of the campaign of MOSOP with that of the Nigerian Conservation Foundation/Hadejia-Nguru Wetlands Conservation Project (NCF/HNWCP) described in the previous chapter is shown in Table 9.2. Both NGOs undertook to campaign on challenging issues, especially in the case of MOSOP. Both NGOs attempted to influence the military government, but they adopted completely different approaches that reflected their very different ideals and membership. Whereas NCF/HNWCP adopted a 'complementary' approach, providing government with policy options aligned with the national interest and backed up with scientific data, MOSOP's 'confrontational' approach made demands that conflicted with the national interest and allowed little room for compromise. MOSOP's campaign gained legitimacy from the support of the Ogoni communities and gained further strength from campaigners in the UK and USA, although this support was not sufficient to protect it from government retaliation. NCF gained legitimacy from HNWCP's scientific expertise and grassroots experience in the wetlands, which helped protect NCF/HNWCP from accusations of being unrepresentative and being driven by the agenda of northern donors.

When the government failed to respond to MOSOP's demands, MOSOP tried to influence the government indirectly by targeting Shell with the help of campaigners in Europe. NCF/HNWCP avoided targeting Berger in Germany in case it antagonised government by attacking Nigeria's international business partners and attracting external intervention by environmental campaigners. NCF/HNWCP avoided overt links with radical campaign groups whereas MOSOP sought them. NCF/HNWCP made only

Table 9.2 Comparison of the campaigns of the NCF/HNWCP (Nigerian Conservation Foundation/Hadejia-Nguru Wetlands Conservation Project) and MOSOP (Movement for the Survival of the Ogoni People) on the basis of (a) contextual factors and (b) the characteristics of the NGO

(a) Contextual factors		
	NCF/HNWCP	MOSOP
i) Nature of the issue	Long-standing development programme of dam construction and irrigation, viewed as important for food self-sufficiency and protection from drought	Oil production in the Niger Delta, crucial to Nigeria's economy (so disruption and demands for self-determination represented extreme challenge to the state)
ii) Stage of policy development (agenda setting, policy choices, or implementation)	Policy choices, but partly with aim of agenda setting (seeking new policies on irrigation)	Right to self-determination, compensation, and prevention of further environmental degradation can be viewed as agenda setting – demanding new policies
iii) Nature of the target organisation	Highly-placed politicians and government officials in key decision-making roles at federal and regional level; (the construction company or its German-based parent company were deliberately not targeted)	Initially the military regime; subsequently Shell (as an international company with its reputation to protect; other oil companies less responsive)
(b) Characteristics of the NGO		
	NCF/HNWCP	MOSOP
i) Legitimacy and capacity	Scientific expertise and evidence; grassroots activities; politically astute and well-connected leaders; international backing from the outset (but not legitimacy through representative membership)	Massive grassroots membership and support; documentary evidence; politically and professionally experienced leaders; international backing later
ii) Strategy (complementary, collaborative, confrontational or consciousness-raising)	Presented as complementary (even though NCF was opposed to government policy)	Confrontation combined with consciousness-raising
iii) Framing the issue	Sustainable use of water resources (not wildlife conservation); in shared national interest; several policy options – room for negotiation	Human rights and justice, predominant over environmental degradation; military government, Shell and ethnic majorities as 'enemies'; all-or-nothing demands and ultimatum
iv) Networking among NGOs	Officially networking only with internationally-known conservation NGOs (WWF, IUCN, RSPB), but unofficial links by some NCF staff with more radical groups (e.g. Greenpeace, CLO, ERA)	Extensive networking, locally and internationally

limited use of the media and appeals to the public, whereas MOSOP exploited them as fully as possible.

The marked contrasts between NCF and MOSOP reflected aspects of the marked contrasts in Nigerian society, in particular the wide gap between the political elite (not only the army but also bureaucracy, business and dominant ethnic groups) and local communities mentioned in Chapter 1. NCF drew from the elite groups, taking advantage of its links with senior politicians, ministry officials, the judiciary and industry to promote its cause. Its membership base was narrow. By contrast, MOSOP relied on extensive support among local communities. NCF was an 'insider' group, unconcerned about adopting a pragmatic approach when necessary to work with the targets of its campaign. MOSOP was an 'outsider' group, attempting to shift the values of government and unwilling to compromise its principles by talking with the 'enemy'.

Both NCF and MOSOP benefited from politically astute and experienced leaders. In particular, both groups were quick to note and respond to any 'window of opportunity' that arose. For example, NCF took advantage of the brief period of the interim civilian government to arrange a meeting with its head, Shonekan, a former NCF member. MOSOP routinely responded immediately to any news event relevant to its cause by writing newspaper articles. Both groups took advantage of longer-term trends and changes. For example, NCF/HNWCP made use of international research showing the economic value of wetlands and international trends away from the funding of large scale dams and irrigation projects. MOSOP took advantage of the general political ferment about the transition to democratic rule in Nigeria to raise the issue of self-determination for minority groups such as the Ogoni. It made use of the heightened international concern about environmental issues and democracy to draw attention to the Ogoni people's plight. Considered together, the two campaigns illustrate how NGOs with very different characteristics and strategies can exert some influence even in the most challenging of circumstances.

References

Amnesty (1996) 'Focus – Nigeria. Just another year', *Amnesty*, Nov/Dec: 16-17.

Ashton-Jones, N. (1996) 'Oil production in the Niger Delta: the economic costs', paper produced for an open forum on *Petroleum and Nigeria's Environment*, convened by the Royal Geographical Society, 15 April 1996.

Douglas, O.N. and Onyeagucha, U. (1995) 'The MOSOP story', unpublished document, prepared for the Open University GECOU project.

Eccleston, B. (1996) 'Does North–South collaboration enhance NGO influence on deforestation policies in Malaysia and Indonesia?', in D. Potter (ed.) *NGOs and Environmental Policies: Asia and Africa*, London: Frank Cass.

Ganesan, A. (1998) 'Partners in crime', *Guardian* (UK), Society Supplement, 20 May: 4.

Lewis, P.M. (1994) 'Endgame in Nigeria? The politics of a failed democratic transition', *African Affairs*, 93: 323–40.

McGreal, C. (1999) 'Day of destiny for Africa's giant', *Guardian* (UK), 27 February: 16.

The Ogonis (1995) 'Our case', *The Masses*, Special Edition: The Death of Ken Saro-Wiwa: 2.

Osaghae, E.E. (1995) 'The Ogoni uprising: oil politics, minority agitation and the future of the Nigerian state', *African Affairs*, 94: 325–44.

PIRC (1996) *Controversies Affecting Shell in Nigeria: Report to Clients*, London: Pensions and Investment Research Consultants Limited, March

Rowell, A. (1995) 'Oil, Shell and Nigeria: Ken Saro-Wiwa calls for a boycott', *The Ecologist,* 25, 6: 210-213.

Salih, M.A. (1999) *Environmental Politics and Liberation in Contemporary Africa*, Dordrecht, the Netherlands: Kluwer

Saro-Wiwa, K. (1995) 'Our case', *The Masses* (Nigeria), 2,1: 1,6.

Shell (1995) *Shell in Nigeria. Shell Briefing Note*, London: Shell International Petroleum Company.

SPDC (1995a) *The Environment. Nigeria Brief*, Lagos, Nigeria: Shell Petroleum Development Company of Nigeria.

SPDC (1995b) *The Ogoni Issue, Nigeria Brief*, Lagos, Nigeria: Shell Petroleum Development Company of Nigeria.

Smith, P. (1998) 'Abubakar begins Nigerian glasnost', *Guardian* (UK), 8 September: 12.

Vidal, J. (1999) 'Shell fights fires as strife flares in the delta', *Guardian* (UK), 15 September: 15.

10 Influence thrust upon them? NGOs' role in public action on the environment in Africa

Alan Thomas, David Humphreys and Susan Carr

We have now analysed seven cases where NGOs contested policies relating to the African environment. The degree of influence NGOs were able to exert varies markedly across the seven cases. What can we conclude? We can make some general points on the reasons why NGOs succeed in influencing policy in some circumstances but not others, following the analytical framework set out in Chapter 1. Thus the chapter will focus first on the political context, then on the policies and, third, on the NGOs. Next we extract from the discussion ten lessons from the cases which NGOs themselves could apply to improve the overall effectiveness of their campaigning strategies. Finally we revisit the question of how NGOs' role should be conceptualised by generalising from the study of particular NGO campaigns to a broader analysis of NGOs' changing role in the search for sustainable development.

Reasons for NGO influence: political context and democracy in Africa revisited

What is it about different political contexts which means NGOs can exert more or less policy influence in them?

Is NGO influence enhanced in countries with a democratic system of government? David Potter defines a democracy as 'a type of political regime in which binding rules and policy decisions are made not by the entire community but by representatives accountable to the community' (Potter 1997: 4). Chapter 1 proposed five features of democracy which might occur in different combinations: free and fair multi-party elections; political and human rights; a plurality of sources of power within and outside the state; a limited reach for the state; and a culture of political participation.

Our case-study countries of Nigeria, Zimbabwe and Botswana exhibit different combinations of these five features. For example, many would see Zimbabwe as effectively an authoritarian one-party state, despite its regular multi-party elections. Our analysis of the domestic political context reveals it to be a country with several democratic features, including a plurality of power centres inside and outside the state and a strong political culture of participation. The state also has a clearly limited reach, its lack of capacity obliging it to welcome both international and national NGOs to pursue a variety of 'gap-filling' activities. Botswana also holds regular multi-party elections, and is often seen as closer to the European liberal democratic model than Zimbabwe. However, the Botswana Democratic Party has won every election since independence in 1966, and there are few competing power centres outside the state. Botswana also has a poorly-developed culture of political participation and the state is involved in almost all areas of economic and social life, co-opting some of the initiatives of NGOs. While Nigeria was under an authoritarian military government during the period of our case studies, some of our five features of democracy were nevertheless in evidence. In particular, the military government did not have the capacity to control all aspects of life, especially locally, and as in Zimbabwe there was space for NGOs to 'fill gaps' and provide expertise. There was also a very lively press, a tradition of public outspokenness and a certain degree of political participation at local level.

The first general conclusion from our case studies in Nigeria, Zimbabwe and Botswana is that NGOs can achieve influence in all three domestic political contexts. There are also cases of failure, to differing degrees, in all three countries. This immediately puts paid to the simple notion that NGOs can only achieve influence in cases where there is a certain minimum degree of democracy – though from the start we eschewed the notion that domestic political contexts can be conceived in terms of 'more' or 'less' democracy. However, it is still arguable that NGO campaigns require the presence of at least some of the features of democracy and the success of different NGO strategies depends on the combination of features present. Below, in the discussion on how NGOs themselves and their strategies can contribute to their own success, we pursue this more complex idea.

However, we should note that, apart possibly from the last, our five features of democracy are based on a European liberal democratic discourse. Hence there are two potential objections to applying them to an African context. First, perhaps liberal democracies are poorly suited to dealing with problems of environmental degradation. Some environmental campaigners do not see democratic processes as of intrinsic value, being concerned only with whether they lead to effective environmental conservation. Andrew Dobson argues that democratic theory concerns itself with means, namely political processes, while green thought concerns itself with ends, namely environmental sustainability. Dobson does not see the two as mutually incompatible, providing democracy concerns itself with ecological outcomes. He notes, however, that in order for such a process to work 'the preconditions need to be so widely drawn (in ecological terms ...) that many modes of decision making (not all of them democratic) can be slipped in' (Dobson 1996: 146). This raises a number of interesting questions, amongst them whether environmental sustainability can be achieved using democratic processes, whether green thought is, as some of its proponents claim, an intrinsically democratic ideology, and whether democracy should be viewed as an end as well as a means and indeed as a central part of environmental sustainability.

Second, many political scientists in Africa are not prepared to accept the European liberal democratic model without qualifications. Concerns with problems such as the economic legacies of colonial rule, underdevelopment, party relations and military–civil relations have been more salient to many African political scientists than democracy (Bangura 1988: 33). Many African scholars have argued that democracy is incomplete unless it embraces questions of equity, economic opportunity and social welfare. For example, and in contradistinction to the liberal democratic view of state legitimation, Adebayo Ninalowo emphasises the importance of socio-political legitimation. Arguing that the social contract between the state and its citizens should provide an 'enabling environment' in which people can meet their basic needs and freedoms, Ninalowo argues that the legitimation of the state 'is most importantly a function of the extent to which it is oriented towards human-centred human development' (1996: 57). To Dei, '[i]f democracy is narrowly defined to exclude food, water, shelter, clothing and employment then the talk about the importance of achieving "sustainable development" for Africa's own good becomes a sham' (1993: 104).

Ninalowo's distinction between state legitimation and socio-political legitimation in Africa is a useful one, although our research reveals that the distinction should not be too widely drawn. One important vehicle for state legitimation is the promotion of policies that acquire socio-political legitimacy in civil society. Certainly state legitimation is an important consideration for the governments of countries that have only relatively recently achieved independence. Most of the NGO campaigners we interviewed appeared sensitive to this and did not wish their campaigns to be seen as a threat to the legitimacy of the state. Indeed in some cases the opposite was true; the campaign against the Kafin Zaki dam succeeded in large part because it was framed in terms of supporting the Nigerian national interest (Chapter 8).

The result is that many NGOs consciously avoid what we have termed the confrontational strategy. They may nonetheless be viewed with suspicion by government. For example, in Botswana there is evidence that some of the more radical NGOs have been viewed by the government as a potential threat to nation-building. One of the claims made by NGOs in Botswana was that the rangelands fencing programme served the interests of the 'cattle barons' and their political allies, rather than the basic needs of the poor (Chapter 6). The Botswana government's response sought to legitimate the policy in socio-political terms; rangelands fencing was intended to serve the interests of all farmers, including the poor. The government was also prepared to shelve the Southern Okavango Integrated

Water Development Project (SOIWDP) when it became clear the project had no socio-political legitimacy, especially among local communities whose leaders claimed their people's livelihoods would be damaged (Chapter 7). In short, African governments should not be conceived as narrowly concerned with state legitimation; the socio-political legitimation of environmental policies is also salient. Meanwhile NGOs are principally concerned with ensuring that environmental policies acquire socio-political legitimacy. Most of them realise that to challenge the legitimacy of the state is generally counter-productive, though it may be a strategy of last resort.

Suspicious of the European model of liberal democracy, many African scholars are also wary of the model of 'good governance' promoted through structural adjustment programmes (SAPs). Some of the criticism of SAPs concerns their impact on living conditions. For example, Mlambo concludes that SAPs increase unemployment and lead to a decline in real wages and 'a serious decline in real incomes and living standards for the populations of the recipient countries in general but particularly for the poor' (1995: 89). However, attempts by the IMF and foreign governments to link the implementation of SAPs to democratic reform are also criticised on the grounds that SAPs are inherently undemocratic. Only a regime with authoritarian tendencies can ensure that SAPs are implemented according to IMF criteria. Hence, with respect to Nigeria, Babawale et al. argue that those financial institutions that promote SAPs display 'a technist concern with democracy' (1996: 119), while Bangura argues that '[r]ather than adjustment promoting democracy, as claimed by its proponents, it, in fact, creates the conditions for a new form of authoritarianism' (1988: 44). Beckman (1991) suggests that it is protest against SAPs, rather than the SAPs themselves, that breeds democracy, and Momoh, noting that the Nigerian military regime suppressed those who protested against the SAP, agrees (1996: 27). Dei argues that, while proposals to empower the African rural poor through the process of democratisation may be laudable, 'we must also be cautious that this becomes an "African democracy", that is, democracy that is understood by African peoples and not something dictated to them by transnational corporations and foreign governments' (1993: 104). Such concerns explain in part the insistence of many African commentators that considerations of democracy must include reference to the economic realm. In the final analysis the economic conditions of SAPs, designed to enable a country better to service its debt to foreign lenders, are more important to international financial institutions than the promotion of liberties through democratic reform. Bluntly put, those who promote SAPs are prepared to sacrifice democracy and environmental sustainability on the altar of global economic stability. International NGOs based in countries that promote SAPs may be regarded with similar suspicion as the promoting institutions themselves.

In our research we have not neglected these criticisms, but neither have we sought to engage in speculation on possible alternatives to the state (for example, Wapner 1995). While liberal democracy may be seen as a model with flaws with respect to policies for ecological sustainability, the case studies in this volume indicate that democratic processes are important as they enable the citizenship of a country to transmit demands for the satisfaction of environmental and human needs to government. Indeed, NGOs are important not only because they may be able to exert influence over policies, but also because their very existence and their activities are potentially part of the democratic workings of their society and may help build up the capacities of civil society to participate in its further democratisation.

Reasons for NGO influence: environmental policies

What is it about certain environmental policies which makes them more or less amenable to influence by NGOs?

Chapter 1 proposed seeking answers to this question by analysing three contextual factors: first the nature of the specific issue with respect to sustainable development and to the interests involved; second the character of the organisations being targeted by the NGOs; and finally the different stages of policy development.

The preliminary discussion in that chapter suggested that policies would be least open to influence where they are most central to the policy-making organisations targeted, and conversely NGO campaigns would be most likely to achieve policy changes in areas relatively peripheral to the target organisations. This argument applied in different ways with all three factors. First, NGO influence would be hardest to achieve where the issues relate closely to important economic or political interests of the target organisations. Second, it would be most difficult to challenge the essential prerogatives of any type of target organisation (for example a government's monopoly over national planning and control of national resources, or a private corporation's responsibility to its shareholders) and easier to achieve influence in areas where a target lacks full capacity to carry out its responsibilities by itself. Third, it would be hardest to influence the central part of the decision process (choices between decision options) and easier to play a part at the start or end of the process (in agenda-setting or implementation).

The INCD case tends to support this view. Thus while large parts of the NGO agenda found their way into the final wording of the Convention to Combat Desertification (CCD – Chapter 3), it is arguable that land degradation is not an issue which affects the central economic or political interests of northern governments, transnational corporations or the big international NGOs. Principles of community empowerment, respect for indigenous knowledge, and a large role for NGOs themselves are all written in to the CCD. However, there may be a tough struggle before these are incorporated in a meaningful way into the results of international negotiations on deforestation or climate change. Indeed agreement of any kind in these areas is endangered by the fact that they challenge the interests of those major actors in a way which agreement on combating desertification and drought does not.

The evidence from our national level case studies suggests that on the whole *direct* influence is indeed easiest in more peripheral policy areas. There were clear-cut changes of government policy following all three water policy campaigns but not in those relating to land use. However, this should not be taken to imply that water is necessarily less central to government interests than land use. Two of the cases involved halting proposed development projects (Chapters 7 and 8). In such cases NGO campaigns can achieve only temporary success. Indeed, water supply in dry areas is likely to become even more of a major policy issue in future decades, and there is always a chance of similar proposals being revived in the name of the 'national interest'. In any event, the interests involved vary considerably from case to case, and one cannot generalise about the differences between 'water use' and 'land use' issues. The water policy issues in this book were each relatively peripheral at the time but for different reasons.

Thus the campaign for water from the Zambezi related to combating drought in certain provinces, not a challenge to any major national economic interests. It succeeded in pushing the pipeline option on to the national planning agenda and in broadening the options for implementation, but not in forcing particular planning decisions. It showed up a specific lack of capacity in government but could not succeed in an overall challenge to government's monopoly control over national resources. The anti-SOIWDP campaign succeeded once it was apparent that the Botswana diamond industry did not need the project; in any case it was not challenging the cattle industry, the other central economic interest in the country. Similarly the campaign to protect the Hadejia-Nguru wetlands was not opposing the Nigerian oil industry or other economic interests of central importance to the Nigerian military.

Conversely the promotion of small grains in Zimbabwe and the argument against rangeland fencing in Botswana affected economic interests with a particular national resonance in each case, namely the maize industry in Zimbabwe and the cattle industry in Botswana. MOSOP's campaign in the Niger Delta challenged both the mighty oil industry and the integrity of the federal state of Nigeria.

However, the lessons from the case studies are not so straightforward as that. In practice many real policy issues are 'central' in some respects and 'peripheral' in others, while influence is not a clear-cut matter of success forcing policy change or else complete failure. Closer inspection of our cases shows some apparent exceptions to the 'more peripheral, easier to influence' rule. Indeed, we went out of our way to study certain cases where influence was achieved apparently against the odds. These exceptions allow us to make some further general points.

First, one way to succeed when challenging a target's central prerogatives is to put forward alternatives that can be absorbed within the existing planning framework. All three water campaigns were challenging the respective governments' prerogatives on national planning, and the anti-SOIWDP and the Hadejia-Nguru campaigns also both contributed to a change of decision, the most central part of the policy process. They were careful not to frame their campaigns in ways which could be portrayed as opposing the national interest but instead made positive proposals for different ways of enhancing that national interest. (The anti-SOIWDP campaign is not such a clear example of this point. Other factors may have been more important in its success, and indeed where the Hadejia-Nguru campaign downplayed the international wildlife angle the anti-SOIWDP campaigners emphasised it, arguing that wildlife tourism could make a big input into the national economy. However, in so doing they were risking a backlash. We return to this below.)

Second, in some circumstances transnational corporations may be easier targets than governments. In particular, TNCs may be sensitive about their international reputations where governments may react strongly against any reference to international environmental responsibilities, possibly seeing these in terms of threats to their sovereignty. So the limited gains which the MOSOP campaign achieved mostly came from targeting Shell. In Botswana, the mere suggestion of the 'Diamonds are for Death' slogan probably played a part in De Beers' early withdrawal of support for the SOIWDP. Both companies could lose substantially in global market terms from any public questioning of their ethical practices. However, targeting TNCs almost certainly implies framing an issue in international terms, and this again could be counter-productive for the NGOs.

Third, it may be possible for NGOs to succeed in shifting future agendas while failing to force any immediate concessions. This was certainly the case with MOSOP's campaign in the Niger Delta. In the case of small grains promotion in Zimbabwe, NGOs may have succeeded in changing the terms of future policy debates by spreading awareness of the small grains option both among communal farmers and among relevant agencies such as Agritex and ICRISAT. To a lesser extent the same may be true of opposition to rangelands fencing in Botswana. In some cases it may become harder for NGOs to win the argument as time goes on, for example if increased water shortages accentuate the apparent need to develop all available water resources despite the risk of irreversible environmental degradation. Environmental consciousness-raising by NGOs may then serve to counter calls for large new development projects and to keep the debate open.

A further point is that there is some force in the suggestion that land use is one of the most difficult areas for NGOs to influence. This is not really demonstrated by our case studies, mainly because we failed to find good cases of NGO influence on the basic questions of land reform, land distribution and land tenure systems. Indeed it was clear from some of our interviews with African NGO activists that they tended to avoid these questions, largely because they constitute one of the most contentious political areas in many African countries, and one where government prerogatives are hardest to challenge.

The case of rangelands fencing in Botswana was one where NGOs failed to make an impact on the policy. Indeed there was no concerted campaign to oppose or modify the government commitment to a programme of continuing enclosure and effective privatisation of land, despite informed misgivings on the part of many Batswana NGOs. This may have been the result of an unspoken recognition that lobbying in this area would very likely have been ineffective, so that any efforts would have wasted scarce NGO resources. MOSOP's campaign was not specifically about land rights but included and went beyond them. It was obliged to adopt a confrontational stance towards government in this most difficult of areas because it saw the environmental and other damage being perpetrated against the Ogoni people they represented as so devastating.

In Zimbabwe the case pursued on small grains promotion was only indirectly about land use. In fact we went to some lengths during 1994 to 1996 to try to find a Zimbabwean case study about the role of NGOs in the debate about the reform of land tenure, but decided that at the time there was no clear story to tell. Like the Batswana NGOs with respect to fencing, Zimbabwean NGOs mostly avoided the issue. Throughout Africa the majority of people are dependent on land for their livelihoods and in southern Africa in particular access to land has been historically unequal, so that one might expect NGOs to be active in advocating more equal access. In Zimbabwe it is the heavily populated communal areas where land degradation is endemic, and where the vicious cycle between poverty and over-exploitation of limited resources of soil, water and forest products tends to appear (Moyo et al. 1991). However, although unequal access to land underlies land degradation, this is not surprisingly seen as a fundamental political issue (Moyo 1995) and only secondarily as an environmental one. Indeed, insistence from NGOs and others on the importance of environmental management in communal areas can be seen as an imposition or a distraction.

There was a review of land tenure policy in Zimbabwe by the Land Tenure Commission (1994) and channels were open for NGOs to make representations to this commission, but few did so. There was some expectation that the commission was likely to reflect the global trend towards privatisation by recommending the widespread adoption of individualised freehold tenure. Concerns about such a move included a fear that it might adversely affect community-based resource management as in the CAMPFIRE schemes, and also the likelihood that it would lead to consolidated land holdings for some and at the same time to large-scale dispossession and hence urban migration of the rural poor. However, in the event, and without any strong NGO input, the commission favoured the continuation of a mixed tenure system, with traditional and communal forms of tenure strengthened alongside a small increase in private freehold tenure.

Since 1994, as moves continue slowly towards implementing some of the reforms proposed by the Land Tenure Commission, certain NGOs have been making more strenuous lobbying efforts on the issue. The Zimbabwean Farmers Union, for example, favours the spread of private landholding. NGOs supporting the CAMPFIRE programme favour strengthening traditional and communal forms of tenure, particularly if that would allow village-level groups legal rights to the exploitation of natural resources, including wildlife. The Zimbabwe Women's Resource Centre and Network lobbies for women to be able to access land in their own right, whether under traditional or individualised tenure. Nevertheless, on the whole it is still the case that the government is proactive in consulting NGOs rather than there being any concerted NGO campaign on the issue. One reason for this is that, although there is agreement on the central importance of land distribution and land tenure to land degradation, as noted in Chapters 1 and 2 there are competing views about the underlying causal mechanisms. There are also a number of different interests involved, several of which are represented via different NGOs, leading to a lack of unity and no clear 'NGO view' on how land tenure should be reformed in order both to ameliorate poverty and to safeguard the environment.

Reasons for NGO influence: NGOs themselves

What is it about particular NGOs and their activities which makes it more or less likely that they will succeed in influencing environmental policies?

Here Chapter 1 suggested considering the legitimacy and capacity of the NGOs, the appropriateness of the strategies they adopt (the 'four Cs'), the relations between them, and finally the way NGOs frame the issue, in order to find explanations for their influence.

The first general conclusion in this area, from the evidence in our case studies, is that there is a tremendous variety of ways in which NGOs achieve influence. Many of the cases involved more than one approach, sometimes in use simultaneously by different NGOs. The legitimacy of different NGOs can rest on a variety of factors. This to some extent relates to the types of NGOs introduced in Chapter 1. For example, the legitimacy of mutual benefit NGOs must depend on the degree to which their membership is representative, while public-benefit research NGOs need to be able to demonstrate expertise. However, as pointed out in Chapter 1, many NGOs are hybrid or combined types and hence legitimacy is sought in complex ways which combine various forms of representativeness with expertise in different fields.

NGO strategies vary partly from playing to their particular strengths in terms of where their legitimacy lies, and also their particular capacities. They also vary in response to different circumstances. We found that the 'four Cs' of collaboration, confrontation, complementary activities and consciousness-raising were useful for categorising the various NGO strategies, but again many of them combined aspects of the four Cs in very particular ways.

The INCD involved many NGOs with very different capacities and varying claims to legitimacy. For example, some such as IIED brought their research expertise to bear via representation on the International Panel of Experts on Desertification. As for participation in the negotiation sessions, it was interesting that accreditation was relatively easy to achieve and, once accredited, all NGOs were treated the same. Some, such as KENGO and ELCI, were experienced in international negotiations while others were attending such sessions for the first time, and the more established NGOs (and the INCD secretariat) helped them out. Thus the NGOs were placed on a steep learning curve, and the progress of the INCD helped them build their capacities for negotiation as well as enhancing their legitimacy in their own countries.

In the three water use cases, some NGOs based their legitimacy on a claim to represent the views of affected populations, while others provided scientific expertise. Thus, in the Zambezi pipeline campaign, MZWP clearly spoke for the ordinary black residents of Bulawayo as well as the white business community and the local political elite, while ZERO and IUCN played important supporting roles when environmental expertise was required. Later, MZWT was deliberately set up by government on the basis of a different kind of representativeness, namely representation of organisational and political stakeholders. In the anti-SOIWDP case, TOCT's membership included both the traditional communities in the Okavango delta area and the wildlife and tourism interests, while the legitimacy of the IUCN derived from its scientific and technical capacity, and the expertise of the mission team they sent to Botswana aimed at an impartial overview without taking sides with particular local interests. By contrast, Greenpeace is a high profile international NGO that is not afraid of 'naming and shaming' tactics, with a global network that can run international campaigns, although they chose not to do so for the Okavango. This status was certainly a source of influence, though from the point of view of the Botswana government it might have not been regarded as 'legitimate' influence. In the Nigerian wetlands campaign, although neither NCF nor HNWCP was based on local membership, legitimacy was gained from the hydrological expertise developed by the project staff and from their grassroots work among wetlands communities.

In the case of the opposition to rangelands fencing in Botswana, none of the NGOs was representative of affected populations and the expertise which they had was not directly in the field of rangelands management. They had low capacity for lobbying work on a policy which, although seen by several as of crucial importance, was not directly in their areas of expertise. To some extent the same comments are true of the NGOs involved in small grains promotion in Zimbabwe. However, in that case one of the NGOs, ORAP, had a strong membership base, while others developed some legitimacy through carrying out their own field trials in a scientific manner and involving local farmers in those trials. The capacity of the Zimbabwe Seeds Action Network (ZSAN) was very limited, and its secretariat, ENDA, was dependent on donor finance. By contrast, in the third of the land use cases the main NGO concerned, MOSOP, had a much higher profile, though its campaign was conducted in terms not of science or technical options but of moral and ethical standards. MOSOP's considerable local legitimacy was countered by the government's regarding it as completely illegitimate due to its claims for local self-determination on behalf of the Ogoni people.

As well as resting on very varied bases, the legitimacy and capacity of the NGOs involved in the different campaigns often changed during the campaigns. As with the INCD, the attempt to influence was itself a source of increased capacity in several cases, as evidence and skills were built up during the course of the campaigns. Thus MZWP deliberately sought evidence about the feasibility of the Zambezi pipeline and how it could avoid environmental damage by commissioning its own studies even before the pipeline was accepted as a strategic option by government. Both NCF and HNWCP in Nigeria and the NGOs involved in promoting small grains in Zimbabwe deliberately used evidence from project work with members of local communities to build up their scientific credibility and their claims to represent local opinion.

Turning to NGO strategies and the four Cs, variety is again the most notable feature. There were two cases of successful confrontation. The campaign against the SOIWDP suggests, to quote Chapter 7, that 'confrontational tactics can work, at least in the short term, in a multi-party democracy like Botswana where there is a plurality of internal power centres and if the tactics are used only in response to occasional crises'. In Zimbabwe, despite that country's less well-established democratic credentials, the Zambezi pipeline campaign was able to trade on the potential for Matabeleland as an alternative power centre. MZWP utilised various electoral campaigns to ensure government support, if only to prevent Ndebele disaffection turning into electoral challenge. MOSOP's campaign in the Niger Delta was also confrontational, but not from choice, and could not be regarded as successful in the short term.

The Nigerian wetlands campaign provides a good example of successful policy influence based on complementary activities. Both the Hadejia Nguru Wetlands Conservation Project itself and at a more general level the environmental expertise provided both by HNWCP and NCF complemented the military government's programmes in fields where its own capacities were lacking. The campaign for water from the Zambezi gives another example, particularly the independent commissioning of studies and local and international fundraising by MZWT. While their success so far is less tangible, the NGOs promoting small grains also used complementary activities to a large extent via the various local projects and field trials which they ran.

While many of our interviewees claimed that collaboration with government was the norm for African NGOs, this strategy does not bring about many clear-cut results. KCS in Botswana, for example, claimed a degree of long-term insider influence. By its nature, it was difficult to obtain hard evidence of the impact of this approach. Certainly the other NGOs who might have opposed the rangelands fencing policy acted collaboratively and had little to show for it. By contrast, NCF in Nigeria was able to use insider connections to help in the campaign to protect the Hadejia-Nguru wetlands, and in Zimbabwe the government-sponsored NGO, MZWT, used a collaborative strategy to press the case for the Zambezi pipeline after MZWP's more confrontational approach had forced the pipeline onto the agenda as an option.

The final C, consciousness-raising, formed a major part of the local campaigns in the cases of the Zambezi pipeline and small grains promotion, as well as for MOSOP both among the Ogoni people themselves and internationally. More broadly, enhanced awareness of environmental issues is one of the major achievements of the environmental movement, but the case studies in this book do not provide much direct evidence of the effects of particular NGOs' activities in this respect.

With regard to our international case study on the INCD, wide ranging consciousness-raising activities were carried out by NGOs in their respective countries. Chapter 3 notes those activities in Nigeria and Zimbabwe. NGOs tended to avoid confrontational tactics at the negotiations. Nonetheless, some delegations did perceive the NGOs as confrontational. When KENGO, speaking on behalf of all the NGOs, raised the issue of land tenure reform, the Kenyan delegation responded that the convention should not interfere with national government policies. However, for the most part NGOs sought to complement the activities of governments, filling in knowledge gaps with respect to their own locality and cooperating with their government delegations. NGOs also sought a complementary role in the implementation of the Convention to Combat Desertification once it was agreed.

Chapter 1 suggested that strategies based on the 'four Cs' may be more or less appropriate depending on the presence or absence of the features of democracy explained above. Evidence from the case studies lends some support to the following hypotheses, which have also been put forward elsewhere (Thomas 1996):

1 NGO influence via the *collaborative*, entryist route may be most effective in a highly participative political culture.

2 A *confrontational* approach will succeed best where there is accountability via free and fair multi-party elections and a plurality of alternative power centres.
3 NGO *complementary* activity may most readily influence policy where the state has a limited reach and there are links to powerful interests external to the country.
4 *Consciousness-raising* will be aided by the upholding of civil and political rights.

However, more research is needed before these hypotheses can be confirmed. The cases in the book already include some exceptions, notably the case of NCF which succeeded in a collaborative approach despite the lack of political participation at national level in a Nigeria run by an authoritarian military regime. It is also clear that the 'hypotheses' cannot simply hold as they stand with any degree of generality, because other factors affect the success of a given strategy apart from the political context.

A stronger conclusion is that the most successful NGO campaigns are *combined* campaigns. Single NGOs may combine more than one basis of legitimacy and adopt a combination of strategies. Most successful campaigns involve several NGOs playing complementary roles.

Thus, in the Zambezi water campaign, MZWP's representative legitimacy was complemented by the expertise of ZERO and IUCN. Its strategy combined confrontation and complementary activities, the latter including commissioning studies and fundraising, both of which in turn improved MZWP's legitimacy in different ways. Similarly, in the Nigerian wetlands campaign, NCF's political acumen was complemented by HNWDP's scientific expertise and legitimacy in representing the interests of local populations gained from its grassroots project work. In this case the strategy combined collaboration with complementary activities, the latter again reinforcing legitimacy. It is interesting to contrast these cases with the promotion of small grains, where the various NGOs concerned largely duplicated each other's activities and indeed were in competition for donor support, or with the example of land tenure reform in Zimbabwe discussed earlier in this chapter, where NGOs represented a variety of interests and promoted opposing positions on the policy.

A final general point here is that NGOs' activities may be aimed at achieving immediate influence, but whether or not they succeed in this respect they will also have effects beyond the immediate. In particular, they may assist NGOs to build up their own capacity, as with several of the examples discussed. This is not always the case, however. For example, TOCT concentrated on local mobilisation as its part of the anti-SOIWDP campaign, and no local capacity was built up from the scientific and technical inputs from Greenpeace and the IUCN team. The way NGOs frame the issue may also have repercussions beyond the immediate, and this relates to an earlier point about how NGOs may succeed in agenda changing for the future while failing to force immediate changes. Thus both the case of small grains promotion and the very different Niger Delta campaign raised general awareness of the issues, making it easier for future campaigners, despite little immediate gain for the respective campaigns themselves. However, the longer-term effects are not necessarily positive for all NGO campaigns. In particular, being obliged to counter NGOs' activities may serve to increase the capacities of governments or other targets, or to strengthen their opposition, as with the Botswana government's tendency to denigrate all local NGO activity as 'white man's business' because of the links with international NGOs such as Greenpeace in cases such as the campaign against the SOIWDP.

Ten lessons for NGO strategy

We can summarise the lessons for NGO strategy from the above discussion:

1. NGOs can achieve influence in very different political contexts.
2. Socio-political legitimation may be more important than the liberal democratic concept of state legitimation for environmental policies in African countries.
3. NGO campaigns are most likely to achieve policy changes in areas relatively peripheral to the target organisations.
4. One way to succeed when challenging a target's central prerogatives is to put forward alternatives that can be absorbed within the existing planning framework.
5. In some circumstances transnational corporations may be easier targets than governments.
6. NGOs may fail to force any immediate concessions, but nonetheless change things for the future, by forcing a new issue onto the agenda, building their own capacity, or provoking a response from the target.
7. Land distribution, land tenure and land reform constitute one of the most difficult areas for NGOs to exert influence. Water use issues may become increasingly contentious as competition for limited resources intensifies.
8. In terms of different bases for legitimacy and different strategies in response to different circumstances, there is a tremendous variety of ways in which NGOs achieve influence.
9. Strategies based on the 'four Cs' of collaboration, confrontation, complementary activities and consciousness-raising may be more or less appropriate depending on the presence or absence of the five features of democracy. Four hypotheses are put forward suggesting when each strategy may work best.
10. The most successful NGO campaigns are *combined* campaigns, with NGOs combining more than one basis of legitimacy and adopting a

combination of strategies, and usually involving several NGOs playing complementary roles.

It is not clear, however, that NGOs could use these ten points directly as a basis for deciding on their strategy. For one thing, many of the factors likely to determine NGO influence are outside their control – for example, where they oppose entrenched political or economic interests. Other factors are not known at the time of a campaign and appear only in retrospect. Nevertheless, together with the more detailed analyses in Chapters 3 to 9 of the particular cases from which the generalisations are drawn, these ten points could help inform some useful analysis of any particular position in which an NGO is developing strategy.

Rethinking NGOs' changing role

As noted in Chapter 1, in the 'New Policy Agenda' (NPA) NGOs are viewed both in economic and in political terms as private actors alongside for-profit firms, and clearly distinguished from public actors, namely the state and its agencies (Edwards and Hulme, 1995). Thus NGOs are seen as private providers of services alongside for-profit firms; and also as one type of private interest group in civil society. This is effectively a two-sector framework (state–private, with NGOs as part of the private sector). However, as argued elsewhere (Thomas 1998, 1999), this framework is inadequate for understanding the place of NGOs in Africa or any other part of the world. The cases in this book reinforce the point that NGOs cannot be regarded simply as private organisations, either in the economic or political sphere. It is also inadequate to consider the economic and political aspects of NGOs' work separately, though for a moment we do so, before putting them together.

The economic aspect of the two-sector approach relies on two important underlying assumptions which are rarely stated explicitly. These are, first, that state agencies are the only ones that act specifically in the public interest so that 'public sector' equates to 'state sector'; and, second, that the economic behaviour of private agencies is basically market-oriented (except where subject to state regulation).

However, the idea that state agencies have a monopoly of the public interest seems particularly inappropriate both in the international sphere and in the case of developing countries where weak states often lack the capacity to provide universal public services or to promote development across the board. Global issues such as environmental degradation involve questions of the public interest while being beyond the capacity of states, even the strongest, to deal with alone. In many, if not most, parts of the world, local communities and particular groups find their collective needs met by neither market nor state.

It is in attempting to meet particular local needs or articulate the interests of particular groups that NGOs, from the local to the international, come into their own. Thus it is surprising that the two-sector framework has been maintained – at least by default – in the thinking of many proponents of the role of NGOs in development. A pro-NGO argument is often put in terms of what is wrong with development by government agencies. Thus Stewart (1997: 12) quotes the following as a good example of 'the anti-government position':

> For government the primary relationship with beneficiaries is one of control, while for NGOs it is one of voluntarism. Governments cannot avoid relating to its [*sic*] citizenry both as policeman and as promoter.
>
> (Fowler and James 1995: 15)

Stewart goes on to point out the overlap between this and neo-liberal orthodoxy, and continues:

> If the state is wholly bad, there is only one alternative and it is private – either private development organisations (NGOs), or the market. This contributes to a very specific definition of NGOs as something to be funded as an alternative to governments or states and furthermore something to be funded because of what they are not (government), rather than because of what they are.
>
> (Stewart 1997: 12)

When we turn to the political sphere, it is again important to emphasise that NGOs are not simply promoting another kind of private interest. Edwards and Hulme (1995: 849) claim that NGOs are 'supposed to act as a counterweight to state power'. However, the idea that NGOs do this by strengthening civil society is quite different from the idea that as private organisations they simply add their lobbying to that of other special interests such as those of business. Van Rooy (1998) suggests that civil society contains organisations that promote public interests as well as those which promote private ones, and points out how it is possible for the former to be marginalised by treating them as though they are only 'special interest groups', like the latter. She cites Barber, who complains how in the USA 'all interests are seen to be particular, each to be weighed as if none represented the common good' (Van Rooy 1998: 17). Thus, according to Barber, environmental groups, '[a]lthough pursuing a genuinely public agenda of clean air for all, including the polluters, ... have been cast as the polluters' twin – another special interest group whose interests are to be arbitrated alongside those of toxic-waste dumpers' (Barber 1996: 272).

Fisher (1998: 13–17) gives detailed examples of six ways in which NGOs promote civil society: first, their proliferation being 'a step towards organizational pluralism' (see also Bebbington and Farrington 1995: 205); second, creating 'vested interests among the poor'; third, promoting political rights and civil liberties (Welch 1995); fourth, '[g]rassroots organizing and bottom-up democratization' which can lead to confronting the state (see also Edwards and Hulme 1992 on 'linking the grassroots with lobbying and advocacy'); fifth, influencing other voluntary organisations; and finally, broadening the ownership of capital through micro-enterprise development. Apart from the last, there are examples of all these in the cases in this book. They are not like lobbying on behalf of the poor as a special interest group; rather, they involve working with other organisations and promoting certain needs of poor communities as aspects of the public interest.

Direct and indirect NGO influence

The cases in this book have mostly concerned attempts at direct influence. In other words, NGOs have campaigned for specific changes in policy. It is rare for NGOs to succeed in obtaining everything they advocate, and it is usual for many outstanding points to remain at issue even after a 'successful' campaign. Nevertheless, in several cases a change can be pinpointed, as with the Botswana government's decision to 'suspend' the SOIWDP (Chapter 7) or the Zimbabwe government's inclusion of the idea of a pipeline from the Zambezi in its national development planning (Chapter 5). In other cases, there has been no set point in time where a clear policy change was made, and it is hard to know whether to regard the NGO efforts as having failed or to argue that there may have been a gradual change in response to NGO activities acting in conjunction with other pressures.

There is a methodological problem here: it is much easier to provide evidence for direct influence leading to specific changes, where it occurs, than to show which indirect influences underlie gradual trends. However, the cases of direct influence may not only be relatively uncommon but also relatively unimportant. Arguably it is often long-term trends which provide the best explanation for policy changes, even where there are immediate causes in terms of direct NGO campaigning. Thus, although in the Zambezi water case, for example, one can point to MZWP's campaign as having forced the issue, it is equally the case that over the period of that campaign, and for some time before and since, there has been a general trend away from rigid government planning and towards including NGOs as well as other elements of civil society and the private sector in a partnership approach to development.

The many examples of this type among our cases studies point to the importance of the interactions between direct and indirect campaigning by NGOs. Of the 'four Cs' introduced in Chapter 1, one (consciousness-raising) refers to indirect influence, while the other three (collaboration, confrontation and complementary activities) are different forms of direct influence.

One of these interactions has been introduced already. Changes in prevailing ways of framing problems may occur over long periods, partly in response to NGO campaigning of the consciousness-raising type, and these changes may then affect how specific issues are settled. One such trend, mentioned above, is that towards a partnership approach involving NGOs. Another is the presumption away from big engineering projects which may have made it easier for the campaigns against the SOIWDP and the Kafin Zaki dam to succeed (Chapters 7 and 8). Yet another is the growing trend since the 1992 UNCED towards participation and the recognition of the importance of indigenous knowledge, written in to the Convention to Combat Desertification (Chapter 3) and giving some backing to the promotion of traditional drought-resistant small grains (Chapter 4). A further example is the consideration now given internationally to the plight of indigenous peoples, which helped MOSOP's campaign (Chapter 9) to gain international support even if it was hard to pinpoint specific gains produced by that campaign. NGOs have long been campaigning internationally for such changes in conceptualising issues, even though it is difficult to show that the activities of a particular NGO or group of NGOs actually explain a particular trend.

Conversely, the direct work of running projects, collaborating over policy or confrontational campaigning all contribute to the legitimacy and capacity of NGOs and hence their effectiveness in undertaking long-term consciousness-raising and their ability to change the terms of debate on key issues. So, for example, the expertise gained by development NGOs in their field-level work on small grains (Chapter 4) allows them to push for relevant policy changes with more authority. We have noted similar arguments for NCF and HNWCP in respect of the Nigerian wetlands campaign. Direct project, policy or campaigning work also generate information which can be analysed and used as evidence in policy debates over a long period. It is not always clear how this will work at the outset. NGOs may painstakingly gather data and engage in low-level consciousness-raising for some time before a 'window of opportunity' occurs and they are able to exploit the legitimacy and evidence which they have built up to influence policy.

One particularly important trend which cannot readily be attributed to direct influence by NGOs is the major movement towards a greater role for NGOs themselves. We noted in Chapter 3 that the impact of the CCD on NGOs was probably greater than that of NGOs on the CCD. There are other examples, at the national level, of NGOs being forced to take on policy roles without independently campaigning for more influence. To paraphrase a well-known quotation: 'Some NGOs are created to be influential. Others have influence thrust upon them.'

However, while NGOs may achieve higher profiles and take on more prominent roles in policy because of a general trend rather than through their own efforts, it is important to realise that they have considerable control over *how* these new roles are conceived. In particular, although they may be given policy roles under the NPA as part of a move towards reducing the power of the state in favour of the private sector, when they actually take up those roles they usually conceive themselves as *public* rather than private actors.

Putting the elements together: a public action approach

Thus whether NGOs in Africa are campaigning on a conservationist agenda or supporting grassroots communities in their search for sustainable livelihoods, they are not promoting private interests but are contesting what is understood as the public interest. Hence we suggest it is more appropriate to apply the concept of 'public action' (Drèze and Sen 1989; Mackintosh 1992) to the activities of NGOs. This means action to meet collective private or perceived public needs, through actions which simultaneously work to redefine what are perceived as public needs. It brings together the economic (meeting public needs) and the political (contesting the perception of public needs), in that meeting needs directly and lobbying for needs to be recognised in the public arena occur together in what is considered to be public action. Thus the following three-point framework is useful for understanding the role of NGOs in global civil society (Thomas 1999):

1 There is a certain political and economic space where needs are unlikely to be fully met by governments or the market, where NGOs may have a comparative advantage.
2 NGOs are best characterised as value-based organisations which are differentiated in terms of which particular values they are based on and which particular groups they focus on or represent.
3 NGOs are part of a range of organisations engaged in public action (alongside various state agencies and international organisations). This means that with others they simultaneously aim at meeting collective private or perceived public needs and help define what are to be regarded as public needs in future.

This last point means that we may be missing something by concentrating on the actions of individual NGOs and how they may or may not influence the policies of other 'target' organisations. Certainly the notion of influence seems to be a legitimate starting point for investigating particular cases, and has been a useful focus for our research. We have tried to explain the success of particular NGOs in meeting public needs through their direct influence on policy. But the concept of public action implies that the cumulative effect of NGOs on the perception of environmental issues in terms of public needs may be at least as important as their immedate exercise of influence.

References

Babawale, T., Fadahunsi, A., Momoh, A. and Olukoshi, A. (1996) 'Nigeria beyond structural adjustment: towards a national popular alternative development strategy', *Africa Development*, XXI, 2 & 3: 119–39.

Bangura, Y. (1988) 'The crisis of underdevelopment and the transition to civil rule: conceptualising the question of democracy in Nigeria', *Africa Development*, XIII: 1, 33–50.

Barber, B. (1996) 'An American civic forum: civil society between market individuals and the political community', *Social Philosophy and Policy*, 13, 1: 269–83.

Bebbington, A. and Farrington, J. (1995) 'Governments, NGOs and agricultural development: perspectives on changing inter-organizational relationships', *Journal of Development Studies*, 29, 2: 199–219.

Beckman B. (1991) 'Empowerment or repression? The World Bank and the politics of African adjustment', *Africa Development*, XVI, 1: 45–72.

Dei, G.J.S. (1993) 'Sustainable development in the African context: revisiting some theoretical and methodological issues', *Africa Development*, XVIII, 2: 97–110.

Drèze, J. and Sen, A. (1989) *Hunger and Public Action*, Oxford: Clarendon Press.

Dobson, A. (1996) 'Democratising green theory: preconditions and principles', in B. Doherty and M. De Greus (eds) *Democracy and Green Political Thought: Sustainability, Rights and Citizenship*, London: Routledge.

Edwards, M. and Hulme, D. (1992) *Making a Difference: NGOs and Development in a Changing World*, London: Earthscan.

Edwards, M. and Hulme, D. (1995) 'NGO performance and accountability in the post-cold war world', *Journal of International Development*, 7, 6: 849–56.

Fisher, J. (1998) *Nongovernments: NGOs and the Political Development of the Third World*, West Hartford, CT: Kumarian Press.

Fowler, A. and James, R. (1995) *The Role of Southern NGOs in Development Cooperation*, Oxford: INTRAC.

Land Tenure Commission (1994) *Report of the Commission of Inquiry into Appropriate Agricultural Land Tenure Systems, Under the Chairmanship of Professor Mandivamba Rukuni*, 3 vols, Harare: Government of Zimbabwe.

Mackintosh, M. (1992) Introduction, in M. Wuyts, M. Mackintosh, and T. Hewitt (eds), *Development Policy and Public Action*, Oxford: Oxford University Press.

Mlambo, A.S. (1995) 'Towards an analysis of IMF structural adjustment programmes in Sub-Saharan Africa (SSA): the case of Zimbabwe 1990–94', *Africa Development*, XX, 2: 77-98.

Momoh, A. (1996) 'The structural adjustment programme and the transition to civil rule in Nigeria (1986–1993)', *Africa Development,* XXI, 1: 19-37.

Moyo, S. (1995) *The Land Question in Zimbabwe*, Harare: SAPES Books.

Moyo, S., Robinson, P., Katerere, Y., Stephenson, S. and Gumbo, G. (1991) *Zimbabwe's Environmental Dilemma: Balancing Resource Inequalities*, Harare: ZERO.

Ninalowo, A. (1996) 'The state, legitimation and human-centred development', *Africa Development*, XXI, 4: 55–73.

Potter, D. (1997) 'Explaining democratization', in D. Potter, D. Goldblatt, M. Kiloh and P. Lewis (eds) *Democratization*, Cambridge: Polity Press/Open University.

Stewart, S. (1997) 'Happy ever after in the market-place: NGOs and uncivil society', *Review of African Political Economy*, 24, 71: 11–34.

Thomas, A. (1996) 'NGO advocacy, democracy and policy development: some examples relating to environmental policies in Zimbabwe and Botswana', *Journal of Commonwealth and Comparative Politics*, XXXIV, 1: 38–65.

Thomas, A. (1998) 'The role of NGOs in development management: a public action approach', *DPP Working Paper 40*, Milton Keynes: Development Policy and Practice, Open University.

Thomas, A. (1999) 'The role of NGOs in global civil society: a public action approach', paper for international conference on *NGOs in a Global Future*, Birmingham, 10–13 January.

Van Rooy, A. (ed.) (1998) *Civil Society and the Aid Industry*, London: Earthscan.

Wapner, P. (1995) 'The state and environmental challenges: a critical exploration of alternatives to the state-system', *Environmental Politics*, 4,1: 44–69.

Welch, C. (1995) *Protecting Human Rights in Africa: Role and Strategies of Non-Governmental Organizations*, Philadelphia: University of Pennsylvania Press.

Annex

Participants in the international seminar on *Environmental Policies and NGO Influence: Land Degradation and Resource Management*, Harare, Zimbabwe, 26–28 June 1996

Hosted by Global Environmental Change Open University (GECOU) and Commutech, Harare

Sponsored by the Economic and Social Research Council (ESRC), The British Council and the Swedish International Development Authority (SIDA)

M. Bolaane	University of Botswana, Gaborone
Andrew Blowers	Open University, UK
Erica Burman	World Resources Institute, Botswana
Susan Carr	Open University, UK
Muyeye Chambwera	ENDA-Zimbabwe, Harare
J. Chidavaenzi	ACRM, Mount Pleasant, Zimbabwe
Priscilla Dembetembe	BIOMASS, Harare
Oronto Douglas	Environmental Rights Action, Nigeria
Kingsley D. Dube	MZWT, Bulawayo, Zimbabwe
Miriam Elderhorst	BOTH Ends, Netherlands
Cosmas Gonese	AZTREC, Masvingo, Zimbabwe
Derek Gunby	Planafric, Bulawayo, Zimbabwe
David Humphreys	Open University, UK
Sylvia Jampies	UN Interim Secretariat for the Convention to Combat Desertification, Geneva
Zephania Kajevu	Development Dialogue, Harare
Badouine Kamatari	Environment Liaison Centre International, Nairobi
Stephen Kasere	CAMPFIRE, Harare
Choga Kudakwashe	CARE International, Harare
Ray Kujeke	ULG Consultants, Zimbabwe
Keith Leggett	Kalahari Conservation Society, Botswana
Lewis Machipisa	Inter Press Service, Harare
Elias Madzudzo	Centre for Applied Social Sciences, University of Zimbabwe

Jestere Manyame	CARE International, Harare
Nelson Marongwe	Institute of Development Studies, Harare
Davison Masendeke	Agritex, Bulawayo, Zimbabwe
Prosper Matonai	Institute of Development Studies, Harare
Jeanlinda Mavunga	ZERO, Harare
Kizito Mazvimavi	Development Technology Centre, University of Zimbabwe
Elizabeth Migonko-Bake	United Nations Environment Programme, Nairobi
K. Moletsane	FONSAG, Botswana
Rob Monro	Zimbabwe Trust, Harare
Patrick Moriarity	Institute of Hydrology, Chiredzi, Zimbabwe
Omphemetse Motumise	University of Botswana, Gaborone
Rodger Mpande	COMMUTECH, Harare
L. Munjoma	Development Dialogue, Zimbabwe
Nontokozo Nabane	Centre for Applied Social Sciences, University of Zimbabwe
Chrispen Ndebele	Matopos Research Station, Bulawayo, Zimbabwe
Thoko Ngwenya	Embassy of Sweden, Harare
Enoch Okpara	Nigerian Environmental Study Team (NEST)
Karen Ross	Conservation International, Maun, Botswana
Onalenna Selolwane	University of Botswana, Gaborone
D. Takavarasha	ENDA-Zimbabwe, Harare
Alan Thomas	Open University, UK

Index